Spectroscopy and Dynamics of Orientationally Structured Adsorbates

WORLD SCIENTIFIC LECTURE AND COURSE NOTES IN CHEMISTRY

Editor-in-charge: S. H. Lin

Vol. 1: Monte Carlo Methods in *Ab Initio* Quantum Chemistry
B. L. Hammond, W. A. Lester, Jr. & P. J. Reynolds

Vol. 2: Quantum Chemistry Aided Design of Organic Polymers: An Introduction to the Quantum Chemistry of Polymers and Its Applications
J.-M. André, J. Delhalle & J.-L. Brédas

Vol. 4: The Physical Chemistry of Biopolymer Solutions: Application of Physical Techniques to the Study of Proteins and Nuclei Acids
R. F. Steiner & L. Garone

Vol. 5: Theoretical Heterogeneous Catalysis
R. A. van Santen

Vol. 6: Density Matrix Method and Femtosecond Processes
S. H. Lin, R. Alden, R. Islampour, H. Ma & A. A. Villaeys

Vol. 7: Spectroscopy and Dynamics of Orientationally Structured Adsorbates
V. M. Rozenbaum & S. H. Lin

World Scientific Lecture and Course Notes in Chemistry – Vol. 7

Spectroscopy and Dynamics of Orientationally Structured Adsorbates

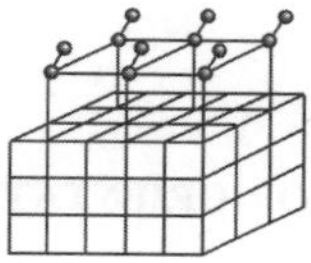

V M Rozenbaum
S H Lin
Academia Sinica, Taiwan

World Scientific
New Jersey • London • Singapore • Hong Kong

Published by

World Scientific Publishing Co. Pte. Ltd.

P O Box 128, Farrer Road, Singapore 912805

USA office: Suite 1B, 1060 Main Street, River Edge, NJ 07661

UK office: 57 Shelton Street, Covent Garden, London WC2H 9HE

British Library Cataloguing-in-Publication Data
A catalogue record for this book is available from the British Library.

SPECTROSCOPY AND DYNAMICS OF ORIENTATIONALLY STRUCTURED ADSORBATES

ISBN 981-238-175-9

This book is printed on acid-free paper.

Printed in Singapore by Mainland Press ac

Contents

Preface

The book covers a variety of questions related to orientational mobility of polar and nonpolar molecules in condensed phases, including orientational states and phase transitions in low-dimensional lattice systems and the theory of molecular vibrations interacting both with each other and with a solid-state heat bath. Special attention is given to simple models which permit analytical solutions and provide a qualitative insight into physical phenomena.

A detailed and rigorous treatment offered by the book is intended mostly for specialists and graduate students in solid state theory and surface physics. It may likewise be attractive for scientists studying the properties of impurity subsystems incorporated in a solid-state bulk or adsorbed on a solid surface, as for instance, intercalated compounds. The researchers in academic institutions or industry who are not strictly surface-oriented but are interested in the subjects mentioned can also take advantage of the material, as a special emphasis on methodological essentials and mathematical tools is provided, thus saving a reader from seeking them elsewhere in the literature. On the other hand, to make the book readable also for neophytes in surface science, the authors tried to keep possibly lucid presentation form throughout the main material (with specific mathematical facets put away into appendices) and to offer a wealth of references on orientational surface structures and vibrational excitations in them. The book is supported by two indices; besides an alphabetical listing of subjects and authors, cross-references will enable a reader to easily access the information both on principal concepts involved and on specific adsorbate compositions.

Chapter 1

Introduction

In the last decade studies on adsorbed molecules and the monolayers formed by them have been receiving a great deal of interest due to diverse features both of applied and fundamental theoretical significance. On the one hand, adsorption and catalysis are extensively used in microelectronics, semiconductor engineering, chemical industry, and biotechnology, which calls for a deeper insight into adsorbate states. On the other hand, a system of adsorbed molecules offers a manifold of facets to be studied: it can be regarded both as an impurity subsystem with respect to a corresponding solid surface and as a quasi-two-dimensional entity in its own right, with all the particular cooperative properties arising from low dimensionality. It is also noteworthy that in most cases an adsorbed molecule is tightly bound to a surface through a single atom, whereas the other atoms (further from the surface) retain considerable mobility. Symmetric arrangement of substrate atoms often admits of several equivalent positions for the atoms of adsorbed molecules, i.e. several equivalent molecular orientations. As a result, the systems concerned are distinguished by high reorientational activity. Provided that intermolecular lateral interactions forming a molecular ensemble are strong enough to remove the orientational state degeneracy for isolated adsorbed molecules, they give rise to orientational ordering. It is evident that changes in temperature may cause phase transitions between orientationally ordered and disordered phases in such systems.

When the concentration of adsorbed molecules is sufficient to form a monolayer, intermolecular lateral interactions become predominating in its orientational structurization and radically change the corresponding spectrum of vibrational excitations, the main source of structural and physicochemical information of the adsorbate. To date a wealth of experimental evidence on adsorbed monolayer structures, relevant monitoring methods, and theoretical approaches has been accumulated and generalized in a number of surveys (see the reference list at the end of Introduction). In view of various factors governing the states of adsorbates, adsorbed monolayers are described, as a rule, on a particular basis, i.e., focusing on a concrete adsorbate/surface combination. Present-day computer-oriented strategies are also adapted to modeling specific systems, with the numerical values of geometrical parameters and interaction constants given. Thus, thorough structural and spectroscopic determinations of concrete character are mainly performed for adsorbates, which meets the current demands of applied science of materials.

There also exists an alternative theoretical approach to the problem of interest which goes back to "precomputer epoch" and consists in the elaboration of simple models permitting analytical solutions based on prevailing factors only. Among weaknesses of such approaches is an a priori impossibility of quantitative-precise reproduction for the characteristics measured. Unlike articles on computer simulation in which vast tables of calculated data are provided and computational tools (most often restricted to standard computational methods) are only mentioned, the articles devoted to analytical models abound with mathematical details seemingly of no value for experimentalists and present few, if any, quantitative results that could be correlated to experimentally obtained data. It is apparently for this reason that interest in theoretical approaches of this kind has waned in recent years.

Straightforward analytical models, however, receive particular attention in the present book, as they are of unique significance in the comprehension of physical phenomena and, moreover, provide the very language to describe them. To exemplify, recall the effect caused on the phase transition theory by the exactly soluble two-dimensional Ising model. Nor can one overestimate the role of the quasiparticle concept in the theory of electronic and vibrational excitations in crystals. As new experimental evidence becomes available, a simplistic physical picture gets complicated until a novel organizing concept is created which covers the facts known from the unified standpoint (thus underlying the aesthetic appeal of science).

The question now arises of what simplification is possible in the treatment of orientationally structured adsorbates and what general model can be involved to rationalize, within a single framework, a diversity of their properties. Intermolecular interactions should include Coulomb, dispersion, and repulsive contributions, and the adsorption potential should depend on the substrate constitution and the nature of adsorbed molecules. However difficult it may seem, all these factors can be taken into account if we follow the description pattern put forward in this book. Its fundamentals are briefly sketched below.

Analyzing orientational structures of adsorbates, assume that the molecular centers of mass are rigidly fixed by an adsorption potential to form a two-dimensional lattice, molecular orientations being either unrestricted (in the limit of a weak angular dependence of the adsorption potential) or reduced to several symmetric (equivalent) directions in the absence of lateral interactions. In turn, lateral interactions should be substantially anisotropic.

The first terms of the power series obtained by the multipole expansion of the Coulomb intermolecular potential account for dipole-dipole interactions prevailing in systems of polar molecules. As an adequate approximation for ensembles of nonpolar molecules with quasinormal or planar orientations of their long molecular axes, we can invoke a generalized form resembling the Hamiltonian of dipole-dipole interactions, with the constant values appropriately renormalized. An additional rationale for such a model follows from the fact that the interactions of dynamic dipole moments of vibrational and electronic transitions represent the basic mechanism for a collectivization of the corresponding excitations. It is therefore expedient to simulate lateral interactions just by a Hamiltonian of a dipole-dipole form, with the interaction constants to be fitted. To account for line shapes in vibrational spectra of adsorbed monolayers, the interaction between the adsorbate and the phonon thermostat of a substrate has also to be involved.

In the present book, we aim at the unified description of ground states and collective excitations in orientationally structured adsorbates based on the theory of two-dimensional dipole systems. Chapter 2 is concerned with the discussion of orientation ordering in the systems of adsorbed molecules. In Section 2.1, we present a concise review on basic experimental evidence to date which demonstrate a variety of structures occurring in two-dimensional molecular lattices on crystalline dielectric substrates and interactions governing this occurrence.

To describe the orientation-dependent part of lateral interactions, we invoke a general angular dependence which is equally relevant to quadrupole, dispersion, and repulsive interactions (all contributing only to distance-dependent coefficients). In the next section (Sec. 2.2), much space is given to the fundamentals of the theory of two-dimensional systems with real (long-range and anisotropic) dipole-dipole interactions. Further (in Sec. 2.3) we demonstrate that the ground states for ensembles of nonpolar molecules with quasinormal or planar orientations can be found by minimization of eigenvalues of the Fourier-transformed short-range dipole-dipole interaction tensor with generalized values of coefficients. This approach affords a straightforward and convenient treatment of the ground states and orientational phase transitions for a system of nonpolar molecules (Secs. 2.4 and 2.5).

Chapter 3 is devoted to dipole dispersion laws for collective excitations on various planar lattices. For several orientationally inequivalent molecules in the unit cell of a two-dimensional lattice, a corresponding number of colective excitation bands arise and hence Davydov-split spectral lines are observed. Constructing the theory for these phenomena, we exemplify it by simple chain-like orientational structures on planar lattices and by the system CO_2/NaCl(100). The latter is characterized by Davydov-split asymmetric stretching vibrations and two bending modes. An analytical theoretical analysis of vibrational frequencies and integrated absorptions for six spectral lines observed in the spectrum of this system provides an excellent agreement between calculated and measured data.

In the pivotal Chapter 4, temperature dependences of spectral line shifts and widths are analyzed. To treat them, we have to invoke rather complex concepts of the theory of many-particle systems. Indeed, temperature dependences of adsorbate spectral lines, with their frequencies strongly exceeding those of atomic vibrations in a substrate, can be accounted for only provided anharmonic coupling with temperature-sensitive low-frequency modes. It is clear that the latter can be represented by low-frequency vibrations of the same adsorbed molecules which should be bound to a thermodynamic reservoir of the substrate and have, in addition, their intrinsic anharmonicity. The problem becomes even more involved at sufficiently high surface concentration of adsorbed molecules, because in this case both high-frequency and low-frequency molecular modes are collectivized due to intermolecular interactions. First we demonstrate that the low-frequency component of the Hamiltonian referring to excitations in the adsorbate and the substrate can be diagonalized in the wave vector of adsorbate excitations which become resonant (i.e. having finite lifetimes) as a result of harmonic coupling with excitations in the substrate (Sec. 4.1). Then, in the framework of the known exchange dephasing model, we introduce a high-frequency mode with biquadratic anharmonic coupling and invoke the Green's function method of the Markov approximation to derive an exact expression for the spectral function of the local high-frequency vibration (Sec. 4.2). In the event that the resonant low-frequency mode is itself strongly anharmonic and can be described by a finite number of harmonic sub-barrier states, the perturbation theory for the Pauli equation is applied so as to account for an additional dependence of spectral line shift and width for a local vibration on a number of sub-barrier states (determined by the reorientation energy barrier). This is followed by the generalization of the exchange dephasing model, first, to various types of anharmonic coupling between high-frequency and low-frequency modes, and second, to collectivized excitations of an adsorbate (Sec. 4.3). Much consideration is given to the contribution of dipolar dispersion laws to dephasing of high-frequency collective vibrations and a simple model for collective high-frequency and low-frequency molecular modes is formulated. The results gained are applied to interpret temperature dependences of spectral line shifts and widths for local vibrations which are observed for a diversity of realistic adsorption systems, *viz.* molecular complexes with hydrogen bonds, OH/SiO_2, H/Si(111), H(D)/C(111), and CO/NaCl(100)).

The book thus embraces an extended study on a variety of issues within the theory of orientational ordering and phase transitions in two-dimensional systems as well as the theory of anharmonic vibrations in low-dimensional crystals and dynamic subsystems interacting with a phonon thermostat. For the sake of readability, the main theoretical approaches involved are either presented in separate sections of the corresponding chapters or thoroughly scrutinized in appendices. The latter contain the basic formulae of the theory of local and resonance states for a system of bound harmonic oscillators (Appendix 1), the theory of thermally activated reorientations and tunnel relaxation of orientational states in a phonon field (Appendix 2), and the temperature Green's function technique in the representation of Matsubara's frequency space (Appendix 3). Thus, practically all the relationships are substantiated in detail mathematically, so that a reader needs not retrieve proofs from other sources. Subject and author indices facilitate a search for a material of interest. To conclude, the book is expected to be helpful as a handbook on the theory of vibrations and reorientations of impurity molecules and groups of atoms and also on the theory of orientational structures and cooperative phenomena in adsorbates.

1. V. M. Rozenbaum, V. M. Ogenko, A. A. Chuiko, *Vibrational and orientational states of surface atomic groups*, Usp. Fiz. Nauk **161**, No 10, 79 (1991) [Sov. Phys. Usp. **34**, 883 (1991)].
2. W. Steele, *Molecular interactions for physical adsorption*, Chem. Rev. **93**, 2355 (1993).
3. D. Marx, H. Wiechert, *Ordering and phase transitions in adsorbed monolayers of diatomic-molecules*, Adv. Chem. Phys. **95**, 213 (1996).
4. H. Ueba, *Vibrational relaxation and pump-probe spectroscopies of adsorbates on solid surfaces*, Progress in Surf. Sci. **55**, 115 (1997).
5. A. Patrykiejew, S. Sokolowski, K. Binder, *Phase transitions in adsorbed layers formed on crystals of square and rectangular surface lattice*, Surf. Sci. Reports **37**, 207 (2000).

Chapter 2

Orientational structures of adsorbates

2.1. Brief survey of experimental data and theoretical approaches

Orientations of long axes of adsorbed molecules are dictated by two factors, *viz.* the angular dependence of the adsorption potential and lateral intermolecular interactions including the screening contribution from a substrate. Strong screening of Coulomb interactions along metal surfaces is likely to result in perpendicular molecular orientations, as for instance in the case of the simplest orientational structures formed by CO molecules on various metal surfaces.[1] Unlike metal substrates, dielectric crystalline adsorbents give rise to a diversity of orientational structures in surface molecular monolayers, as long axes of adsorbed molecules are allowed to deflect from the surface-normal direction and their orientation projections onto the surface plane are completely governed by longitudinal lateral interactions (see surveys [2-4] and references cited therein).

Structures of this kind are normally detected by neutron, low-energy electron, and X-ray diffractions. The systems under study can be exemplified by the monolayers N_2,[5-8] O_2[9] (Figs. 2.1, 2.2), CO_2,[10] C_2N_2,[11] CS_2[12] on a graphite surface (long molecular axes lie in the surface plane) or by the monolayer of CO_2/NaCl(100)[13-19] (molecules are slightly surface-inclined) (Fig. 2.3). Each of these orientational configurations can be treated as a two-sublattice monolayer, with the parameters given in Fig. 2.4 and the corresponding values listed in Table 2.1.

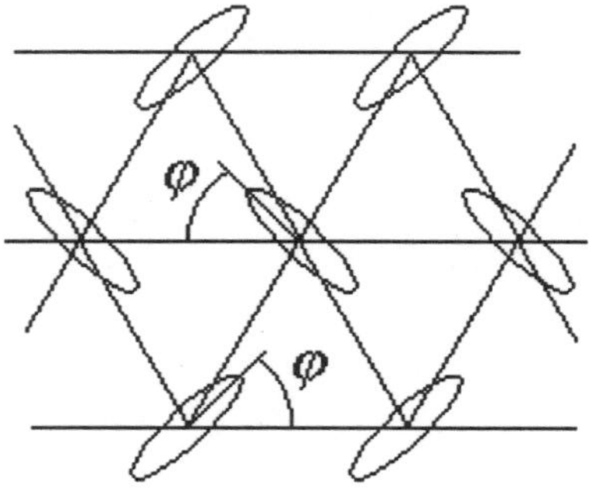

Fig. 2.1. The molecular orientations in the commensurate $\sqrt{3}\times\sqrt{3}$ N_2 monolayer on the graphite basal plane.

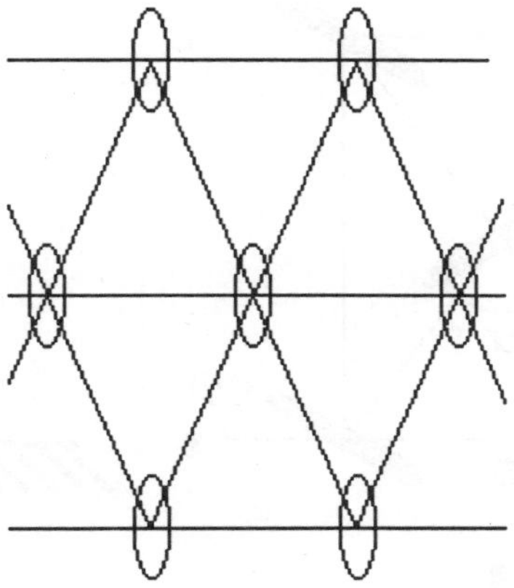

Fig. 2.2. The molecular orientations in the ordered O_2 monolayer on the graphite basal plane.

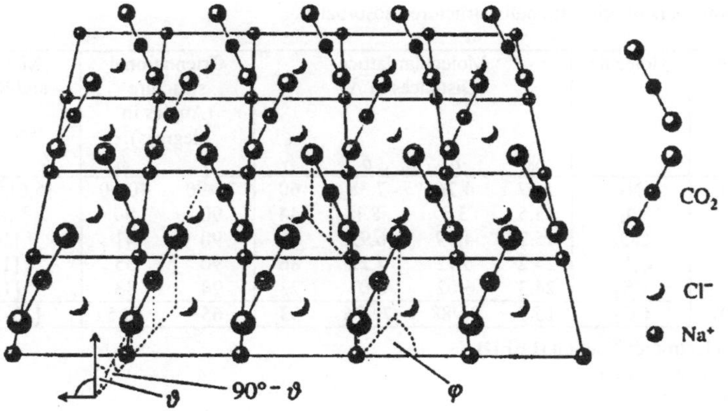

Fig. 2.3. (2x1)-structure of CO_2/NaCl(100) at monolayer coverage.[15-17]

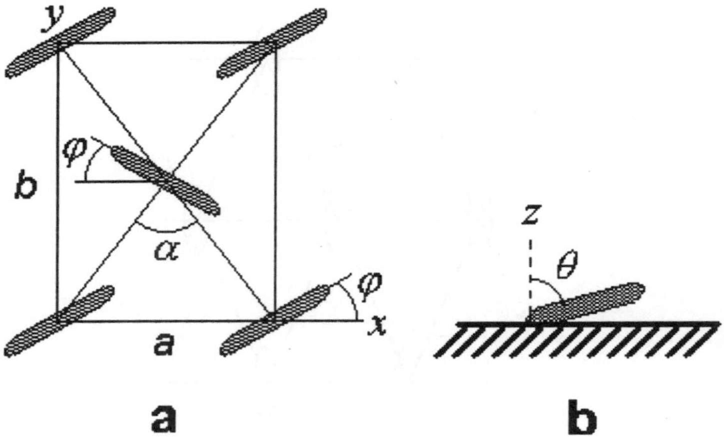

Fig. 2.4. Definition of structural parameters for a two-sublattice monolayer: (a) projections of long molecular axes onto the unit cell; (b) side view showing the tilt of the molecular axis.

Table 2.1. Parameters of orientationally structured adsorbates

Substrate	Molecule	Molecular lattice (Distances in Å)				Orientational structure (Angles in degrees)		Method and Refs.
		S_0	a	b	α	θ	φ	
Graphite	N_2	15.7	4.26	7.38	60	75-90	40-50	[a] [5,6] [b] [7,8]
	O_2	13.5	3.3	8.1	44	90	90	[a] [9]
	CO_2	15.2	4.39	6.93	65	90	41	[c] [10]
	C_2N_2	24.2	6.72	7.22	86	90	35	[b] [11]
	CS_2	24.3	6.02	8.07	73	90	33	[c] [12]
NaCl(100)	CO_2	15.9	3.988	7.976	53	65	49.5	[13-19]

[a] Low-energy electron diffraction (LEED)
[b] Elastic neutron diffraction
[c] X-ray diffraction

Projections of molecular axes onto the surface plane form chain-like structures in which the chains with identically oriented molecules alternate (with the exception of oxygen molecules). The Davydov splitting of spectral lines represents the main spectroscopic manifestation of adsorbed structures with several orientationally inequivalent molecules in the unit cell of a two-dimensional adsorbate lattice. Many adsorbates, such as CO_2/NaCl(100),[13-17,20-26] CO/NaCl(100),[20,27-32] N_2O/NaCl(100),[33] CO_2,SO_2/CsF(100),[34] HN_3/NaCl(100),[35,36] CO_2/MgO(100),[26,37,38] CO/MgO(100)[38,39] etc. belong to this class. As an example, the monolayer CO_2 on NaCl(100) forms a (2×1) structure with pg symmetry, containing two molecules per adsorbate unit cell (Fig. 2.3). Infrared absorption of CO_2 molecules was observed in the region of the asymmetric stretching vibration ν_3 (~ 2345 cm^{-1}) and the bending mode ν_2 (~ 665 cm^{-1}). The asymmetric stretching vibration appears as a Davydov doublet, the symmetric collective vibrational mode being inclined with respect to the surface, the antisymmetric mode being parallel to the surface (Fig. 2.5).

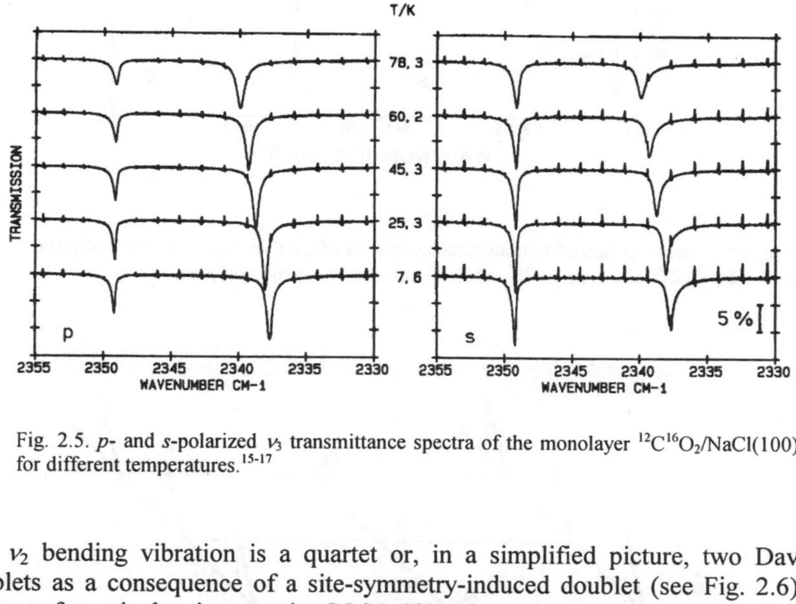

Fig. 2.5. *p*- and *s*-polarized ν_3 transmittance spectra of the monolayer $^{12}C^{16}O_2$/NaCl(100) for different temperatures.[15-17]

The ν_2 bending vibration is a quartet or, in a simplified picture, two Davydov doublets as a consequence of a site-symmetry-induced doublet (see Fig. 2.6).[40] A system of particular interest is CO/NaCl(100): it is characterized by inclined molecular orientations with θ=25^0 and antiferroelectric ordering of chains at low temperatures (see Fig. 2.7) which is removed on the phase transition at $T \approx 25$ K. This structural information is deduced from the observed Davydov splitting of the spectral line for the CO stretching vibrations at 2155 cm^{-1} and T<24 K (see Fig. 2.8).[20,27,28,41-45]

Theoretical treatment of such structures is favored by the fact that the azimuthal component of the angular dependence of the adsorption potential is determined by the arrangement symmetry for the substrate atoms closest to the adsorbate, i.e., by the substrate crystal lattice. As a result, an isolated molecule can have several degenerate equiprobable orientations in the surface plane and lateral interactions in the ensemble of adsorbed molecules remove the degeneracy thus dictating the

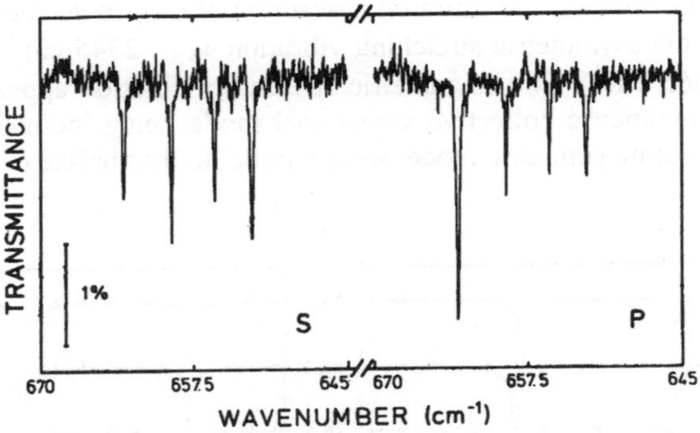

Fig. 2.6. p- and s-polarized ν_2 transmittance spectra of the monolayer $^{12}C^{16}O_2$/NaCl(100) at $T = 80$ K (CO_2 pressure 1×10^{-9} mbar, instrumental resolution 0.05 cm^{-1}).[40]

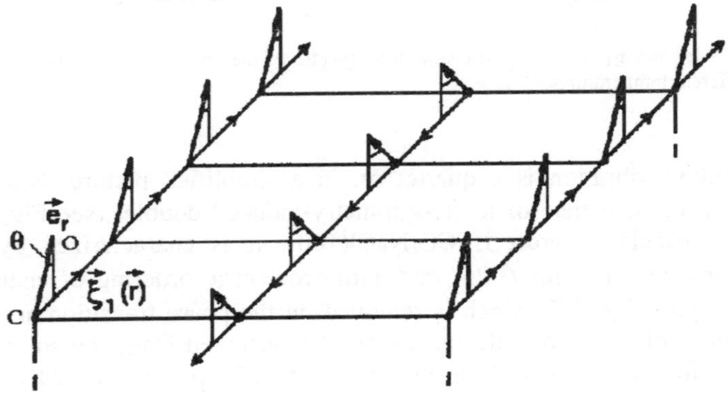

Fig. 2.7. (2x1)-structure of CO/NaCl(100) at monolayer coverage.

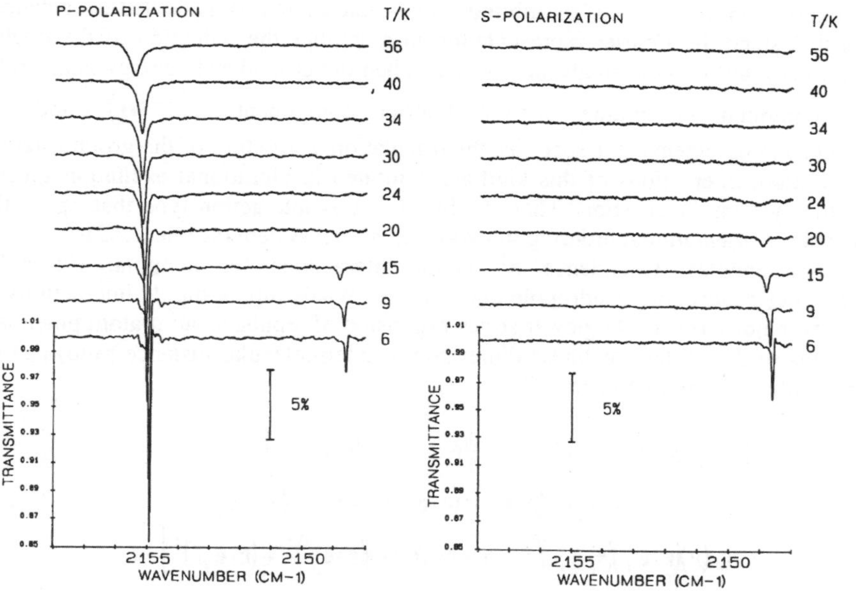

Fig. 2.8. Monolayer transmittance spectra of the adsorbate $^{12}C^{16}O/NaCl(100)$ for different temperatures. Angle of incidence 50±1°.[27,28]

orientational structure of the system. Lateral interactions include Coulomb, dispersion, and repulsive interactions, the first of them predominating at distances comparable with the separation between neighboring adsorbed molecules.

For polar molecules, dipole-dipole interactions provide a paramount contribution to Coulomb interactions. The orientational structure is determined by the ground state of the dipole-dipole Hamiltonian:

$$H_D = \frac{1}{2}\sum_{j\neq j'} V^{\alpha\beta}\left(\mathbf{r}_{jj'}\right) e_j^{\alpha} e_{j'}^{\beta}, \quad V^{\alpha\beta}(\mathbf{r}) = D_1\delta^{\alpha\beta} + D_2 \hat{r}^{\alpha}\hat{r}^{\beta} \tag{2.1.1}$$

(here $D_1 = \mu^2/a^3$, $D_2 = -3\mu^2/a^3$, μ is the dipole moment, a is the lattice constant, $\hat{\mathbf{e}}_j$ denotes the orientation unit vector for the jth molecule, $\hat{\mathbf{r}} \equiv \mathbf{r}/r$ is the unit vector directed from one molecular center to another, and summation over twice-repeated

Greek indices $\alpha, \beta = x, y, z$ of the Cartesian coordinate axes is meant). Minimization of expression (2.1.1) for an arbitrary two-dimensional Bravais lattice becomes possible since the Fourier representation in $\mathbf{q}$ implies the reduction of the double sum over j and j' to the single sum over $\mathbf{q}$. Then the ground state energy is given by the minimum eigenvalue of the Fourier component, $\widetilde{V}^{\alpha\beta}(\mathbf{q})$, and the corresponding eigenvector specifies the orientational structure of the ground state.[46] In addition, interactions of this kind are exhibited in vibrational excitation spectra for the systems in question. That is why, it is this interaction type that opens the analysis of orientational ordering in two-dimensional system (see Sec. 2.2).

The angular dependence of lateral interactions for nonpolar molecules (including quadrupole-quadrupole and Van der Waals dipole-dipole interactions as well as major terms of the power series expansion of repulsive atom-atom potentials in terms of the molecular linear dimension to intermolecular distance ratio) can be represented in a unified form:[47-52]

$$\begin{aligned} V\left(\mathbf{r}, \hat{\mathbf{e}}_j, \hat{\mathbf{e}}_{j'}\right) = B_0(r) + B_1(r)\left[\left(\hat{\mathbf{r}}\cdot\hat{\mathbf{e}}_j\right)^2 + \left(\hat{\mathbf{r}}\cdot\hat{\mathbf{e}}_{j'}\right)^2\right] \\ + B_2(r)\left(\hat{\mathbf{r}}\cdot\hat{\mathbf{e}}_j\right)^2\left(\hat{\mathbf{r}}\cdot\hat{\mathbf{e}}_{j'}\right)^2 + B_3(r)\left(\hat{\mathbf{e}}_j\cdot\hat{\mathbf{e}}_{j'}\right)^2 \\ + B_4(r)\left(\hat{\mathbf{r}}\cdot\hat{\mathbf{e}}_j\right)\left(\hat{\mathbf{r}}\cdot\hat{\mathbf{e}}_{j'}\right)\left(\hat{\mathbf{e}}_j\cdot\hat{\mathbf{e}}_{j'}\right) + B_5(r)\left[\left(\hat{\mathbf{r}}\cdot\hat{\mathbf{e}}_j\right)^4 + \left(\hat{\mathbf{r}}\cdot\hat{\mathbf{e}}_{j'}\right)^4\right] \end{aligned} \quad (2.1.2)$$

in which the coefficients $B_0(r)$-$B_5(r)$ (listed in Table 2.2 of Sec. 2.3) depend only on the absolute magnitude (modulus) of the intermolecular distance r. Potential (2.1.2) contains quartic terms in $\hat{\mathbf{e}}_j$ and enables no simplification of the problem on switching to the Fourier representation. In the particular case of quasi-normal orientations of long molecular axes relative to the surface plane, the sum of pairwise interactions (2.1.2) on planar symmetric lattices of adsorbed molecules can be minimized analytically by reducing it to the dipole Hamiltonian (2.1.1) with the renormalized coefficients D_1 and D_2. In another particular case of a planar system of nonpolar molecules, the double-angle technique enables lateral interactions to be considered likewise in terms of the dipole-dipole Hamiltonian with renormalized interaction constants, which provides great advance in the elucidation of some significant issues. First, the ground state structures can be determined for nonpolar molecules on square and triangular lattices at arbitrary values D_1 and D_2 and hence at arbitrary relationships between quadrupole, dispersion, and repulsive interactions. The corresponding phase diagrams constructed in section 2.3 directly show which ground states are characteristic of molecular ensembles with certain values of the constants D_1 and D_2. Second, an exact thermodynamic description of the orientation phase transition can be provided on the basis of the two-dimensional Ising and isotropic Ashkin-Teller models subject to the condition that a molecule has two or four discrete orientations in the surface plane (this requirement can be satisfied by a certain angular dependence of the adsorption potential). This method is invoked in sections 2.4 and 2.5 to calculate phase transition temperatures on square and triangular lattices of planar quadrupoles. Experimentally observed orientational structures of adsorbates and the corresponding phase transitions are rationalized in the light of the approach developed.

2.2. Orientational ordering in two-dimensional dipole systems

The first step in studying the orientation ordering of two-dimensional dipole systems consists in the analysis of the ground state. If the orientation of rigid dipoles is described by two-dimensional unit vectors $\mathbf{e_r}$ lying in the lattice plane, then the ground state corresponds to the minimum of the system Hamiltonian

$$H = \frac{1}{2}\sum_{\mathbf{r},\mathbf{r}'} V^{\alpha\beta}(\mathbf{r}-\mathbf{r}')e_{\mathbf{r}}^{\alpha}e_{\mathbf{r}'}^{\beta} \tag{2.2.1}$$

in which $\mathbf{r} = n_1\mathbf{a}_1 + n_2\mathbf{a}_2$ $(n_1, n_2 = 0, \pm 1, ..., \; a_1 \le a_2)$ are the sites of a two-dimensional Bravais lattice, and the tensor $V^{\alpha\beta}(\mathbf{r})$, in the case of dipole-dipole interaction, is defined by the equation

$$V^{\alpha\beta}(\mathbf{r}) = VD^{\alpha\beta}\left(\frac{\mathbf{r}}{a_1}\right), \quad V = \frac{\mu^2}{a_1^3}, \quad D^{\alpha\beta}(\mathbf{r}) = \frac{\delta_{\alpha\beta}}{r^3} - 3\frac{r_\alpha r_\beta}{r^5} \tag{2.2.2}$$

(μ is the dipole moment). The twice repeated Greek superscripts $\alpha, \beta = x, y$ of the Cartesian coordinate axes indicate summation. The total energy and the periodic structure of the orientations of N dipoles in the ground state are defined by the equations:[46,53]

$$\begin{gathered} H = \frac{1}{2}N\tilde{\tilde{V}}_1, \quad \mathbf{e_r} = \vec{\xi}_1(\mathbf{r}), \\ \vec{\xi}_1(\mathbf{r}) = \sum_{l=1}^{L} C_{jl}\tilde{\vec{\xi}}_1(\mathbf{k}_l)\exp(i\mathbf{k}_l \cdot \mathbf{r}), \\ \sum_{l=1}^{L}\left|C_{jl}\right|^2 = 1, \; j = 1, \; 2. \end{gathered} \tag{2.2.3}$$

Here $\tilde{V}_1 = \min \tilde{V}_1(\mathbf{k})$; $\tilde{V}_j(\mathbf{k})$ (with $\tilde{V}_1(\mathbf{k}) \le \tilde{V}_2(\mathbf{k})$) and $\tilde{\xi}_j(\mathbf{k})$ are the eigenvalues and eigenvectors of the Fourier components of the dipole interaction tensor:

$$\tilde{V}^{\alpha\beta}(\mathbf{k}) = \frac{1}{2}\sum_{\mathbf{r}} V^{\alpha\beta}(\mathbf{r})\cos(\mathbf{k}\cdot\mathbf{r}),$$
$$\tilde{V}^{\alpha\beta}\tilde{\xi}_j^{\beta}(\mathbf{k}) = \tilde{V}_j(\mathbf{k})\tilde{\xi}_j^{\alpha}(\mathbf{k}), \quad \tilde{\tilde{\xi}}_j(\mathbf{k})\cdot\tilde{\tilde{\xi}}_{j'}(-\mathbf{k}) = \delta_{jj'} . \tag{2.2.4}$$

The "star" of the L wave vectors $\mathbf{k}_i$ corresponds to the degenerate (at $L > 1$) minimum eigenvalue $\tilde{V}_1(\mathbf{k}_i)$.

The unit length of all vectors $\tilde{\xi}_1(\mathbf{r})$ for an arbitrary anisotropic interaction significantly limits the possible periodic configurations of dipole moments in the ground state, which, except for multidomain structures, may be only homogeneous, and also with a double or quadruple period of the lattice: k_i, = $\mathbf{h}/2$, $\mathbf{h}/4$, where $\mathbf{h} = h_1\mathbf{b}_1 + h_2\mathbf{b}_2$ is an arbitrary vector of the reciprocal lattice[46] ($\mathbf{a}_j\mathbf{b}_j = 2\pi\delta_{ij}$, $h_1, h_2 = 0, + 1,...$). For lattice systems with dipole-dipole interaction the arbitrary minima $\tilde{V}_1(\mathbf{k})$ with $\mathbf{k} \neq h/2$ are absent and the search for configurations of dipoles in the ground state is sufficiently limited by values $\mathbf{k}$=0, $\mathbf{b}_1/2$, $\mathbf{b}_2/2$, $(\mathbf{b}_1 + \mathbf{b}_2)/2$. This corresponds to minimization of Eq. (2.2.1) with respect to the orientations of $\mathbf{e}_r$, in four sublattices examined in the two-dimensional analog[54] of the Luttinger-Tisza method.[55] The quantities $\tilde{V}_1(\mathbf{k})$, generally speaking, cannot be identical to the four listed vectors $\mathbf{k} = \mathbf{h}/2$. Only two cases are possible: when one of these quantities is minimal, or (for a lattice with symmetry axes above the second order) the two values are equal: $\tilde{V}_1(\mathbf{b}_1/2) = \tilde{V}_1(\mathbf{b}_2/2)$. These considerations show that the Luttinger-Tisza method is burdened by independent minimization variables, while analysis of the values of the Fourier components $\tilde{V}_1(\mathbf{k})$ makes it possible to immediately exclude no less than half of the variable set and to obtain a result much more quickly. Degeneracy of the ground state occurs either due to coincidence of minimal values of $\tilde{V}_1(\mathbf{k})$ at two boundary points of the first Brillouin zone $\mathbf{k} = \mathbf{b}_1/2$ and $\mathbf{k} = \mathbf{b}_2/2$, or as a result of the equality $\tilde{V}_1(\mathbf{k}) = \tilde{V}_2(\mathbf{k})$ at the same point $\mathbf{k} = \mathbf{h}/2$. The natural consequence of the ground state degeneracy is the presence of a Goldstone mode in the spectrum of orientational vibrations.[53]

Let us present the ground state characteristics of dipoles (interacting as defined by Eq. (2.2.2)) on square, triangular, rectangular, and rhombic lattices. The ground state of a square dipole lattice was first determined by the Luttinger-Tisza method in

Ref. 54, and by minimization of $\widetilde{V}_1(\mathbf{k})$ in Ref. 56. The energy of this state, if calculated per dipole, is $H_0 = (1/2)\widetilde{V}_1(\mathbf{b}_1/2) = (1/2)\widetilde{V}_1(\mathbf{b}_2/2) = -2.5495\,V$, which is about $\Delta H_0 = 0.291\,V$ lower than the energy of the ferroelectric state. The corresponding configurations of dipoles have a microvortex structure with a period of $2a$ and with the degeneracy in angle α (Fig. 2.9a).

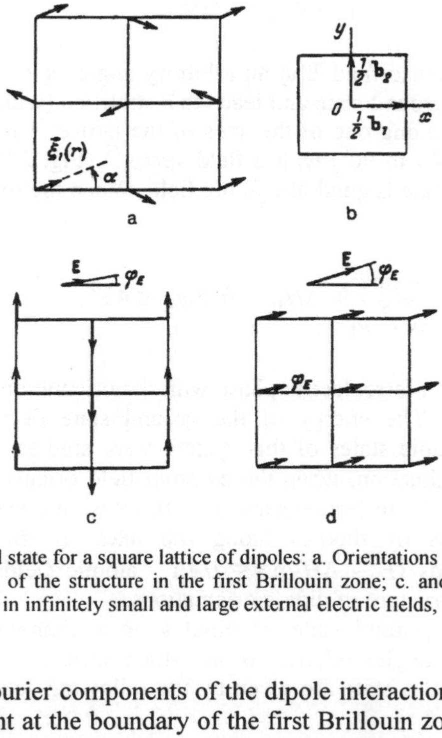

Fig. 2.9. Ground state for a square lattice of dipoles: a. Orientations of dipole moments; b. wave vectors of the structure in the first Brillouin zone; c. and d. orientations of dipole moments in infinitely small and large external electric fields, respectively.

The expansion of Fourier components of the dipole interaction tensor in the vicinity of the minimum point at the boundary of the first Brillouin zone, with the Cartesian axes Ox and Oy respectively chosen along $\mathbf{b}_1$ and $\mathbf{b}_2$ (see Fig. 2.9b), has the form

$$\widetilde{D}^{\alpha\beta}(\mathbf{k}+\mathbf{b}_1/2) = \begin{pmatrix} 6.033 & 0 \\ 0 & -5.099 \end{pmatrix} + \begin{pmatrix} -1.170 & 0 \\ 0 & 0.145 \end{pmatrix} q_x^2 + \begin{pmatrix} -1.333 & 0 \\ 0 & 1.786 \end{pmatrix} q_y^2 - 0.879 \begin{pmatrix} 0 & 1 \\ 1 & 0 \end{pmatrix} 2q_x q_y\,, \tag{2.2.5}$$

where $q = ka \ll 1$, $a = a_1 = a_2$, and the expression for $\widetilde{D}^{\alpha\beta}(\mathbf{k}+\mathbf{b}_2/2)$ follows from Eq. (2.2.5) with the substitutions $q_x \leftrightarrow q_y$ and $xx \leftrightarrow yy$ of matrix components. The eigenvectors in Eq. (2.2.4) and the coefficients C_{jl}, in Eq. (2.2.3) are defined as follows:

$$\tilde{\vec{\xi}}_2(\mathbf{b}_1/2) = \tilde{\vec{\xi}}_1(\mathbf{b}_2/2) = (1,\ 0),\quad \tilde{\vec{\xi}}_2(\mathbf{b}_2/2) = \tilde{\vec{\xi}}_1(\mathbf{b}_1/2) = (0,\ 1),$$
$$C_{11} = -C_{21} = \sin\alpha\ ,\quad C_{12} = C_{22} = \cos\alpha\ . \tag{2.2.6}$$

An infinitely small electric field E at an arbitrary angle $\varphi_\mathbf{E}$ to the Ox axis removes the degeneracy of the ground state and leads to a stratified (with period $2a$) structure of dipole orientations along one of the axes of the lattice. This is the largest angle (on the interval from 45 to 90^0) with a field vector[57] (Fig. 2.9c). In this phase the energy of the ground state is quadratic in the field. When the following approximate equation is satisfied

$$\mu E \approx 2\frac{1-\sin\varphi_\mathbf{E}}{\cos^2\varphi_\mathbf{E}}\Delta H_0,\quad 0 \le \varphi_\mathbf{E} \le 45^0, \tag{2.2.7}$$

the system switches to a ferroelectric phase with the orientation of the dipoles along the field (Fig. 2.9d). The energy of the ground state decreases linearly as E increases. The metastable states of this system were studied in Ref. 58. Thus, in contrast to the usual situation, when the external field orients the dipole moments along itself (for example, in ferroelectrics), in this case the perturbing field causes transverse orientations of dipoles along the axes of the lattice. Moreover, sufficiently weak fields ($E \sim \Delta H_0/\mu \ll H_0/\mu \sim$ atomic fields) may regulate the abrupt change in the structure of dipole orientations.

The ferroelectric ground state of dipoles on a triangular lattice, with the degenerate inclination angles relative to the lattice axes, was revealed in Ref. 56. The characteristics of this state are given by the following equations (see also Ref. 59):

$$H_0 = \frac{1}{2}\widetilde{V}_1(0) = \frac{1}{2}\widetilde{V}_2(0) = -2.7585\ V,$$
$$\widetilde{D}_1(k) \approx -\widetilde{D}_0 + 0.2633\ q^2,\quad \widetilde{D}_2(k) \approx -\widetilde{D}_0 + (4\pi/\sqrt{3})q, \tag{2.2.8}$$
$$\widetilde{D}_0 = 5.517,\quad \tilde{\vec{\xi}}_1(0) = (\cos\alpha,\ \sin\alpha),\quad \tilde{\vec{\xi}}_2(0) = (-\sin\alpha,\ \cos\alpha),$$
$$C_{11} = C_{21} = 1,$$

where $q = ka \ll 1$. The full dependences $\tilde{D}_j(\mathbf{k})$ along symmetrical directions of the first Brillouin zone for square and triangular dipole lattices were calculated in Ref. 59 and are presented in Fig. 2.10.

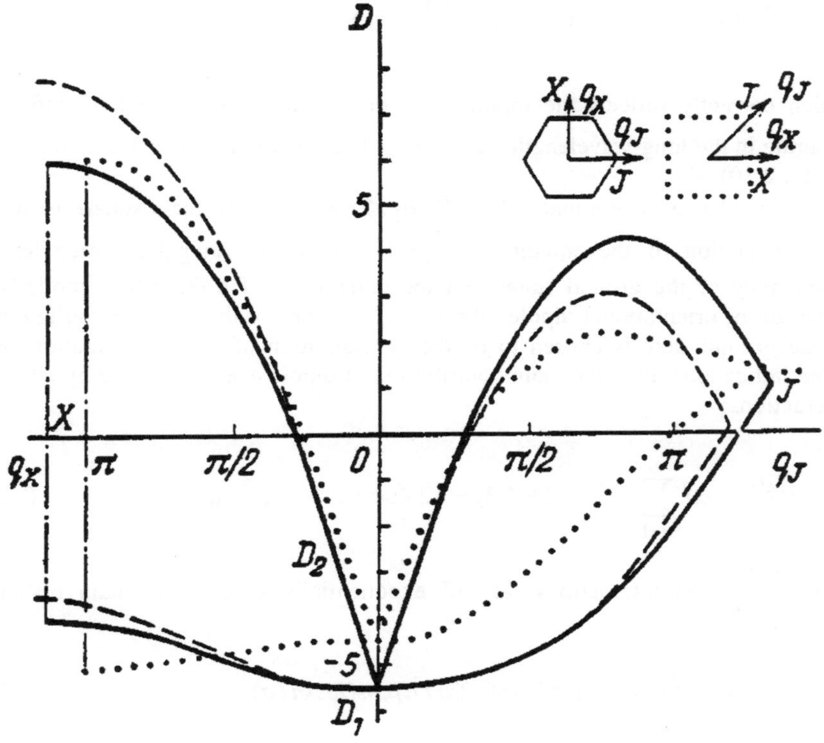

Fig. 2.10. Eigenvalues of the Fourier component of the dipole-dipole interaction tensor in two-dimensional infinite lattices. The solid lines are for a triangular lattice, the dashed lines are for an analytical approximation (2.2.9), and the dotted lines are for a square lattice.

In an approximation of the circular (radius $q_m = 3.733$) first Brillouin zone of a triangular lattice, the following analytical expression was obtained:[59]

$$\tilde{D}_1(q,\Phi) = -D_0\,(1 - \frac{24}{35}\tilde{q}^2 - \frac{12}{35}\tilde{q}^4 \cos 6\Phi)$$

$$\tilde{q} = \frac{q}{q_m},\ \Phi \text{ – angle } (\mathbf{q}, Ox) \tag{2.2.9}$$

$$\tilde{D}_2(q,\Phi) = -D_0\,[(1 - \frac{24}{5}\tilde{q} + \frac{108}{35}\tilde{q}^2 (1 + \frac{2}{9}\cos 6\Phi)]$$

which correctly reflects the topology of the surfaces $\tilde{D}_j(\mathbf{q})$ and is sufficiently accurate in the long-wavelength region $q << 1$ (compare the dashed and solid curves in Fig. 2.10).

Coincidence of minimal values $\tilde{V}_1(\mathbf{b}_1/2) = \tilde{V}_1(\mathbf{b}_2/2)$ for a square lattice and the satisfaction of the equality $\tilde{V}_1(0) = \tilde{V}_2(0)$ for a triangular lattice leads to degeneracy of the ground states and the emergence of a Goldstone mode in the spectrum of orientational dipole vibrations.[46,53] If one of the possible configurations of the ground state is chosen to be the configuration of dipole orientations along some lattice axis, then the main contribution to dipole energy is made by intrachain interactions:

$$H_0^{\text{ch}} = -2V\sum_{n=1}^{\infty} n^{-3} = -2V\zeta(3) \approx -2.404\ V,\quad V = \mu^2/a^3 \tag{2.2.10}$$

while interchain interactions fall off exponentially as the interchain distance z increases:[59]

$$H_0^{\text{int}} \approx [8\pi^2 V/(z/a)^{1/2}]\exp(-2\pi z/a)\cos(2\pi\Delta a/a),\quad z \geq a, \tag{2.2.11}$$

where a is the distance between chain sites, Δa is the shift in sites in neighboring chains. For a triangular lattice, $z = (\sqrt{3}/2)a$, $\Delta a = a/2$ and the negative value $H_0^{\text{int}} = -0.354\ V$ indicates the ferroelectric ordering of dipoles in neighboring chains. For a square lattice, $z = a$, $\Delta a = 0$ and we obtain the positive value $H_0^{\text{int}} = 0.146\ V$ which changes its sign at antiferroelectric ordering of dipoles in neighboring chains (corresponding to the ground state).

Two-dimensional Bravais lattices with no higher than second-order axes of symmetry are characterized by a non-degenerate dipole ground state. On a rectangular lattice, the dipoles are oriented along the chains with the least intersite distances a_1 and antiferroelectric ordering in neighboring chains. As an example, for the rectangular lattice with $a_2 = \sqrt{3}a_1$, the ground state energy is defined, accurate to 0.1%, by intrachain interaction (see Eq. (2.2.10)), because $H_0^{\text{int}} = 10^{-3}\,V$.

The ground states of rhombic lattices with an arbitrary rhombic angle α were studied using the Luttinger-Tisza method in Ref. 60. A description of these states using a chain distribution of interactions (such as Eqs. (2.2.10) and (2.2.11)) was presented in Ref. 61. Figure 2.11 presents different configurations of dipoles which satisfy conditions of periodicity with $\mathbf{k} = \mathbf{h}/2$. The dependences of the corresponding dipole energies H_0/V on the rhombic angle α are given in Fig. 2.12.

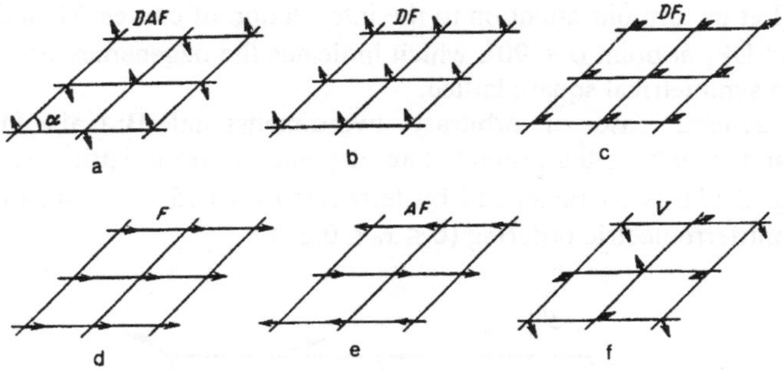

Fig. 2.11. Configurations of dipole moments in a two-dimensional rhombic lattice.

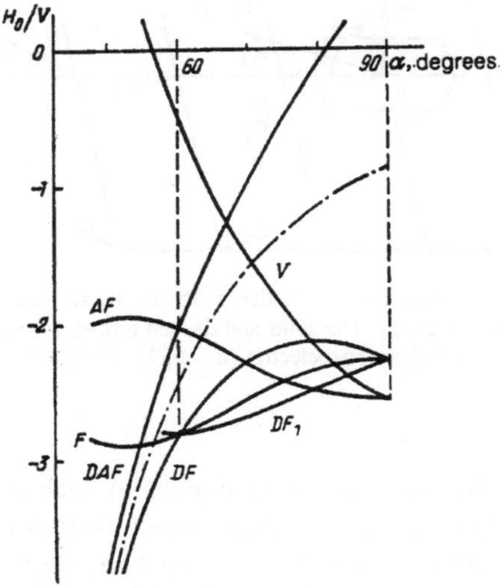

Fig. 2.12. Dependences of the dipole energies of various orientational states (see Fig. 2.11) on the rhombic angle α.

At $\alpha \leq 60^0$ the ground state is a diagonal ferroelectric phase DF with orientation of the dipoles along the minor diagonals of the rhombi (Fig. 2.11b). The axis line between the phase curves DF and DAF corresponds to the contribution of intrachain interactions $-\zeta(3)/[4\sin^3(\alpha/2)]$. For a triangular lattice ($\alpha = 60^0$), the energies of the phases DF, F, and DF_1 coincide, so that the parallel orientations of the dipole moments may form an arbitrary angle with the axes of the lattice. In the range $60^0 \leq \alpha \leq 80^0$, the ground state is also ferroelectric, but with the dipole moments oriented along the major diagonals of the rhombi (phase DF_1 in Fig. 2.11c). Finally, at $80^0 \leq \alpha \leq 90^0$, the antiferroelectric phase AF is favorable in terms of energy (see Fig. 2.11d). Let us turn our attention to the intersection of curves AF and V, as well as DF, F, and DF_1 at point $\alpha = 90^0$, which indicates the degeneracy of these groups of states for a symmetrical square lattice.

In the general case of arbitrary two-dimensional Bravais lattices (not rectangular and rhombic), the ground state, depending on the lattice parameters (x_0 and y_0 in Fig. 2.13), is characterized by ferroelectric ($0.25 \leq x_0 \leq 0.5$) or stratified bisublattice antiferroelectric ordering ($0 \leq x_0 \leq 0.25$).

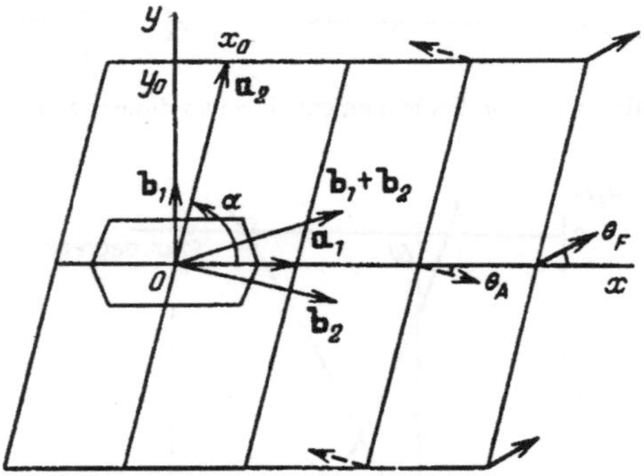

Fig. 2.13. Two-dimensional Bravais lattice with the basis vectors $\mathbf{a}_1$, $\mathbf{a}_2$, and the reciprocal lattice vectors $\mathbf{b}_1$, $\mathbf{b}_2$. The solid and dashed arrows at angles θ_F and θ_A give the ferroelectric ($\mathbf{k} = 0$) and antiferroelectric ($\mathbf{k} = \mathbf{b}_1/2$) configurations of dipoles in the ground state.

The parallel or antiparallel orientations of dipoles to each other may form certain angles θ_F or θ_A with the Ox axis of the lattice, which is drawn along the chains with the smallest intersite distances[46] ($a_1 \leq a_2 \leq |\mathbf{a}_1 - \mathbf{a}_2| \leq |\mathbf{a}_1 + \mathbf{a}_2|$). The energies of these ground states can also be conveniently calculated in the chain representation with somewhat more awkward[46] expressions than Eqs. (2.2.10) and (2.2.11).

Thus, in a study on the properties of dipole systems, most promise is shown by the representation of chain interactions, which, first, reflects the tendency toward ordering of dipole moments along the axes of chains with a small interchain to intrachain interaction ratio. Second, this type of representation makes it possible to use, with great accuracy, analytical equations summing the interactions of all the dipoles on the lattice. Third, there are grounds for the use of the generalized approximation of an interchain self-consistent field presented in Refs. 62 and 63 to describe the orientational phase transitions.

Ferroelectric ordering in certain infinite two-dimensional lattices is due to the long-range contribution of dipole forces. Thus, it is not surprising that in limited two-dimensional lattices numerical calculations of dipole interactions lead to the replacement of ferroelectric states with macrovortex states[64] which approximate to ferroelectric states far from the center of the limited lattice (coinciding with the center of the macrovortex).

Analysis of the ground states of two-dimensional dipole systems would be a purely methodological problem with no practical value if the thermodynamic fluctuations which arise at infinitely small (but nonzero) temperatures could disrupt the orientational ordering. This situation arises, for example, in a one-dimensional Ising model[65] with the interaction $V(r) \propto r^{-3}$. Proof of the ground state stability and the existence of long-range order at low temperatures in two-dimensional systems with dipole interaction and arbitrary (degenerate) dipole orientations in the lattice plane was presented in Refs. 66 and 67.

Resolve the unit vectors of dipole orientation $\mathbf{e}_\mathbf{r}$ into the vectors $\xi_1(\mathbf{r})$ and $\xi_2(\mathbf{r})$ of the ground state, which are defined by Eq. (2.2.3)

$$\mathbf{e}_\mathbf{r} = \xi_1(r)\cos\varphi_\mathbf{r} + \xi_2(r)\sin\varphi_\mathbf{r} . \tag{2.2.12}$$

Then the angles φ_r can be regarded as fluctuations with respect to the ground state. At low temperatures, φ_r are small and the Gaussian approximation of long-wavelength fluctuations of the angular Fourier components $\tilde{\varphi}(\mathbf{k})$ permits the Hamiltonian of Eq. (2.2.1) to be written as:

$$\mathrm{H} = \frac{1}{2}N\tilde{V}_1 + \frac{1}{2}\sum_\mathbf{k} J(\mathbf{k})|\tilde{\varphi}(\mathbf{k})|^2 , \tag{2.2.13}$$

where

$$J(\mathbf{k}) = \sum_{l=1}^{2} C_{2l}^{2} \tilde{V}^{\alpha\beta}(\mathbf{k} + \mathbf{h}_l/2)\xi_2^{\alpha}(\mathbf{h}_l/2)\xi_2^{\beta}(\mathbf{h}_l/2) - \tilde{V}_1 . \quad (2.2.14)$$

In this approximation, the parameter of long-range order η acquires the form

$$\eta^2 = \lim_{r\to\infty} \left\langle \cos(\varphi_{\mathbf{r}_1+\mathbf{r}} - \varphi_{\mathbf{r}_1}) \right\rangle = \exp\left(-\frac{T}{N} \sum_{\mathbf{k}} \frac{1}{J(\mathbf{k})} \right) \quad (2.2.15)$$

and the presence of long-range order ($\eta \neq 0$) depends on the convergence of sums over **k** of $J^{-1}(\mathbf{k})$.

Anisotropic dipole interaction in two-dimensional lattices with a no higher than the second-order symmetry axis leads to the fact that the function $J(\mathbf{k})$ does not go to zero at any **k**. Actually, in this case $l = 1$, $\tilde{V}_2(\mathbf{h}_1/2) > \tilde{V}_1(\mathbf{h}_1/2)$ in Eq. (2.2.14) and

$$J(\mathbf{k}) = \tilde{V}_1(\mathbf{k} + \mathbf{h}_1/2)\sin^2\alpha_{\mathbf{k}} + \tilde{V}_2(\mathbf{k} + \mathbf{h}_1/2)\cos^2\alpha_{\mathbf{k}} - \tilde{V}_1(\mathbf{h}_1/2), \quad (2.2.16)$$

where $\alpha_{\mathbf{k}}$ is the angle between the vectors $\xi_2(\mathbf{k} + \mathbf{h}_1/2)$ and $\xi_2(\mathbf{h}_1/2)$. At $\mathbf{k}=0$, we have $\alpha_{\mathbf{k}}=0$ and $J(0) = \tilde{V}_2(\mathbf{h}_1/2) - \tilde{V}_1(\mathbf{h}_1/2) > 0$. At $\mathbf{k}\neq 0$, due to the inequality $\tilde{V}_2(\mathbf{k} + \mathbf{h}_1/2) \geq \tilde{V}_1(\mathbf{k} + \mathbf{h}_1/2) > \tilde{V}_1(\mathbf{h}_1/2)$, we also have $J(\mathbf{k}) > 0$. Thus, at the sites of the examined lattices, the dipole interaction forms local potentials (proportional to $(1/2)J_{\min}\varphi_{\mathbf{r}}^2$ at small $\varphi_{\mathbf{r}}$) which stabilize the long-range order. Indeed, to substantiate the long-range order in such systems, one need not examine the thermodynamics of the jump-like changes (by large angles) in dipole orientations between the local equilibrium positions, since the limiting case of the situation concerned, viz. an exactly solved two-dimensional Ising model with short-range interaction, implies long-range ordering.

If a ferroelectric ground state is realized on a two-dimensional lattice with a symmetry axis of order higher than two (as with a triangular dipole lattice), the long-wavelength asymptotic behavior of the tensor $\tilde{V}^{\alpha\beta}(\mathbf{k})$ can be written in the form (see Eq. (2.2.8)):

$$\tilde{V}_{\alpha\beta}(\mathbf{k}) = \tilde{V}_1(k)\delta_{\alpha\beta} + (\tilde{V}_2(k) - \tilde{V}_1(k))k_\alpha k_\beta k^{-2} \quad (2.2.17)$$

and, according to Eq. (2.2.14),

$$J(\mathbf{k}) = \tilde{V}_1(k) - \tilde{V}_1 + (\tilde{V}_2(k) - \tilde{V}_1(k))(\mathbf{k}\cdot\tilde{\xi}_2(0))^2 k^{-2}. \qquad (2.2.18)$$

In the approximation of a circular Brillouin zone, the desired sum over **k** in Eq. (2.2.15), if substituted by the integral over the angle between **k** and, acquires the form[66]

$$\frac{1}{N}\sum_{\mathbf{k}}\frac{1}{J(\mathbf{k})} \approx \frac{S_0}{2\pi}\int_0^{k_m}\frac{kdk}{\{[\tilde{V}_1(k) - \tilde{V}_1][(\tilde{V}_2(k) - \tilde{V}_1]\}^{1/2}} \qquad (2.2.19)$$

(S_0 is the area of the unit cell). An analogous integral arose in Ref. 61 for the correlator component transverse to an external electric field **E**:

$$\left\langle \varphi_r^2 \right\rangle \sim \int_0^{k_m}\frac{kdk}{\{[(\mu E/\eta) + \tilde{V}_1(k) - \tilde{V}_1][(\mu E/\eta) + \tilde{V}_2(k) - \tilde{V}_1]\}^{1/2}}. \qquad (2.2.20)$$

For isotropic short-range interactions, we have $\tilde{V}_1(k) = \tilde{V}_2(k) \approx \tilde{V}_1 + \gamma k^2$ and Eq. (2.2.20) switches to the well-known equation[68]

$$\left\langle \varphi_r^2 \right\rangle \propto \int_0^{k_m}\frac{kdk}{(\mu E/\eta) + \gamma k^2} \qquad (2.2.21)$$

thus proving the absence of long-range order in a two-dimensional system with short-range Heisenberg interaction[69] (since it is only at $\eta \to 0$ that the divergence arising at $E \to 0$ can be eliminated for the integral in $\left\langle \varphi_r^2 \right\rangle$ which is bounded by definition). For $E \to 0$ and $\eta \neq 0$, the integral in Eq. (2.2.20) coincides with Eq. (2.2.19), and its convergence signifies the presence of long-range order.

For a triangular dipole lattice, the long-wavelength asymptotic behavior of $\tilde{V}_1(k)$ and $\tilde{V}_2(k)$ is defined by Eq. (2.2.8). The linear dependence of $\tilde{V}_2(k)$ arising from the substitution of Eq. (2.2.8) into Eq. (2.2.19) causes the denominator of the integrand to be proportional to $k^{3/2}$; as a result, the integral converges, and the ferroelectric ground state is stable $(\eta \neq 0)$.[66,67]

In the case of a square dipole lattice, one can substitute Eqs. (2.2.5) and (2.2.6) into the general equation (2.2.14) and obtain, at T=0, the following dispersion law $J(\mathbf{k})$ of orientational vibrations:

$$J(\mathbf{k}) \approx V[1.786(q_x^2 \sin^2 \alpha + q_y^2 \cos^2 \alpha) + 0.145(q_x^2 \cos^2 \alpha + q_y^2 \sin^2 \alpha)]\,. \quad (2.2.22)$$

It can be shown that the thermodynamic fluctuations renormalize this dispersion so that the new function $\mathcal{J}(\mathbf{k}) \approx \text{const} \cdot T + J(\mathbf{k})$ differs from zero for all wave vectors $\mathbf{k}$ and $T \neq 0$. This destroys the Goldstone mode and gives rise to long-range order with the corresponding parameter η^2 (see Eq. (2.2.15)) proportional to exp(-$T|\ln T|$).[70] The mechanism of this renormalization of the dispersion law is related to the well-known ordering effect of thermodynamic fluctuations on intersublattice antiferromagnetic orientations of the magnetic moments which were degenerate in the ground state.[71,72] In contrast to the ground state energy H, the dispersion law $J(\mathbf{k})$ and the free energy of the system

$$F(T \to 0) = H - \frac{T}{2}\sum_{\mathbf{k}} \ln \frac{2\pi T}{J(\mathbf{k})} \quad (2.2.23)$$

depend on the degeneracy parameter α. Minimization of (2.2.23) with respect to α suggests that discrete symmetry with collinear orientations of the dipole moments in the sublattices ($\alpha = 0$ or $\pi/2$) emerges.

Ferromagnetic ordering of two-dimensional systems with dipole-dipole and exchange interactions in the approximation of a spherical model was examined in Ref. 73. The main simplifying assumption of the spherical model is the replacement of the condition $|\mathbf{e}_\mathbf{r}| = 1$ on the orientation vectors with the weaker condition $\sum_{\mathbf{r}} \mathbf{e}_\mathbf{r}^2 = N$, which leads to the incorrect conclusion that the long-range order is absent in two-dimensional dipole systems with a continuously degenerate ferromagnetic ground state (the error was eliminated in Ref. 67).

Based on an examination of the partition function involving the dipole orientations in the functional of electric fields, and taking into account only the long-wavelength asymptotic behavior $k_\alpha k_\beta / k^2$ of the dipole interaction tensor on a triangular lattice (which corresponds to a continuous approximation), it was also erroneously inferred that long-range order was absent in arbitrary two-dimensional dipole systems.[74] To refute that standpoint, recall that first, such systems are characterized by another long-wavelength asymptotic expression of the type $k_\alpha k_\beta / k$ and second, the ground state in these systems can correspond to the wave vectors $\mathbf{k}$ at the boundary of the first Brillouin zone, so that in spite of all the elegance of the field functional formalism for dipole systems, one cannot get around the discrete nature of the lattice.[67]

The presence of local potentials such as hindered-rotation potentials in addition to dipole interactions of Eq. (2.2.1) immediately stabilizes long-range order.[61,66,67] Taking into consideration only nearest-neighbor dipole interactions, the limiting case of two or four discrete orientations of dipoles in the lattice plane makes it possible to reduce the problem of calculating the statistical sum to an exactly solvable two-dimensional Ising model.[75] To exemplify, for a square lattice with two allowed orientations of dipoles along some axis of the lattice, or four orientations along the bisectors of angles between the axes ($\alpha = 0$ and $\pi/4$ in Fig. 2.9a), the transition temperatures will be $T_c = 3.282$ or $1.641\ V$ respectively (the difference between the two values by a factor of two accounts for the transition from a one-dimensional to two-dimensional orientation space).

The description of phase transitions in a two-dimensional dipole system with exact inclusion of long-range dipole interaction and the arbitrary barriers ΔU_φ of local potentials was presented in Ref. 56 in the self-consistent-field approximation. The characteristics of these transitions were found to be dependent on ΔU_φ and the number n of local potential wells. At n=2, T_c varies from $|\tilde{V}_1|/2$ to $|\tilde{V}_1|$ as ΔU_φ increases from 0 to ∞. At n=3, the transition to the ferroelectric phase is a first-order transition (in contrast to cases with any other n implying second-order transitions). As ΔU_φ varies from 0 to ∞, the transition heat q and temperature T_c vary as follows: $q = 0$ to $|\tilde{V}_1|/8$, $T_c = (|\tilde{V}_1|/2)$ to $(3|\tilde{V}_1|/8\ln 2)$. However, for $n \geq 4$ and for arbitrary ΔU_φ, when $T_c = |\tilde{V}_1|/2$, the dependence on ΔU_φ is still preserved in the coefficient of η^4 in the Landau expansion at $n = 4$.

It is well known that the self-consistent field approximation overestimates the transition temperatures. The previously estimated[66,67] lower limit of T_c for a triangular dipole lattice with degenerate orientations ($\Delta U_\varphi = 0$): $T_c > 0.693\ V$ is about four times less than the corresponding estimate provided by the self-consistent field approximation, $T_c = 2.7585\ V$. The lower phase transition temperatures in two-dimensional dipole systems may be explained using the chain representation of interactions (see Eqs. (2.2.10) and (2.2.11)). The strong intrachain dipole interaction cannot ensure long-range order in an isolated chain and low phase transition temperatures are thus caused by the small value of interchain interactions. Good estimates of transition temperatures T_c which consider the small value of interchain interactions are obtainable using a generalized approximation of an interchain self-consistent field[62,63] which involves an exact solution of a one-dimensional Ising model with short-range interaction $H_0^{\mathrm{ch}} = -2V$. The quantity T_c

is a factor of 2 smaller for two-dimensional dipoles than for one-dimensional dipoles; therefore, according to Refs. 62 and 63, we arrive at:

$$T_c \approx \frac{2V}{\ln\left|H_0^{\text{ch}} / H_0^{\text{int}}\right|} . \tag{2.2.24}$$

Substituting values of H_0^{int} in the above expression, we obtain really low values $T_c \approx 1.15V$ and $0.76V$ for triangular and square lattices respectively.[67] The Monte Carlo method was used to evaluate the phase transition temperature $T_c \approx 0.75V$ for a square dipole lattice[76]. On the previously studied[77] rectangular dipole lattice with $a_2 = \sqrt{3}a_1$, the phase transition occurs at $T_c \approx 0.26V$, according to Eqs. (2.2.11) and (2.2.24). With the parameters taken as in Ref. 77, $\mu = 2.5$ D and $a_1 = 5$ Å, we arrive at the estimate $T_c \approx 94$ K. The asymptotic behavior of T_c on the lattices with large interchain distances ($z >> a$) is directly deducible[67] from Eqs. (2.2.11) and (2.2.24):

$$T_c \approx \frac{1}{\pi}\frac{\mu^2}{a^2 z} . \tag{2.2.25}$$

Full consideration of intrachain interactions (beyond the nearest-neighbor approximation) and interchain interactions is included in the self-consistent field and leads to an increase in the transition temperatures in Eqs. (2.2.24) and (2.2.25) by a factor[46] of $\zeta(2) \approx 1.645$.

2.3. Lateral interactions of nonpolar molecules and expressing them in a quasi-dipole form

The orientational contribution to the Hamiltonian for adsorbed nonpolar molecules can be represented in the most general form as a sum of local adsorption potentials $U(\theta_j, \varphi_j)$ and lateral intermolecular interactions $V^{\alpha_1\alpha_2\alpha_3\alpha_4}(\mathbf{r}_{jj'})$, the latter involving quadrupole-quadrupole and Van der Waals dipole-dipole (dispersion) interactions as well as the dominant terms of a series got by expanding the repulsion potential in powers of the molecular linear dimension to intermolecular distance ratio:

$$H_{\rm or} = \sum_j \left[U(\theta_j, \varphi_j) + e_j^{\alpha_1} e_j^{\alpha_2} e_j^{\alpha_3} e_j^{\alpha_4} \sum_{j'(\neq j)} \tilde{V}^{\alpha_1\alpha_2\alpha_3\alpha_4}(\mathbf{r}_{jj'}) \right] + \frac{1}{2} \sum_{j \neq j'} V^{\alpha_1\alpha_2\alpha_3\alpha_4}(\mathbf{r}_{jj'}) e_j^{\alpha_1} e_j^{\alpha_2} e_{j'}^{\alpha_3} e_{j'}^{\alpha_4}. \qquad (2.3.1)$$

The summation over j and j' refers to all positions of centers of mass $\mathbf{r}_j$ for adsorbed molecules. The unit vectors

$$\mathbf{e}_j = (\sin\theta_j \cos\varphi_j, \sin\theta_j \sin\varphi_j, \cos\theta_j) \qquad (2.3.2)$$

specify molecular orientations defined by the angles θ_j and φ_j of the spherical coordinate system, with the axis Z perpendicular to the surface plane and the summation performed over twice-used indices $\alpha_1,\ldots,\alpha_4 = x,y,z$ of the corresponding Cartesian axes. Based on previously reported results,[47-49,52,78,79] the lateral interaction tensors, $V^{\alpha_1\alpha_2\alpha_3\alpha_4}(\mathbf{r}_{jj'})$ and $\tilde{V}^{\alpha_1\alpha_2\alpha_3\alpha_4}(\mathbf{r}_{jj'})$, can be written as:

$$\tilde{V}^{\alpha_1\alpha_2\alpha_3\alpha_4}(\mathbf{r}) = \frac{1}{2} B_0 \delta^{\alpha_1\alpha_2} \delta^{\alpha_3\alpha_4} + B_1 r^{\alpha_1} r^{\alpha_2} \delta^{\alpha_3\alpha_4} + B_5 r^{\alpha_1} r^{\alpha_2} r^{\alpha_3} r^{\alpha_4}, \qquad (2.3.3)$$

$$V^{\alpha_1\alpha_2\alpha_3\alpha_4}(\mathbf{r}) = B_2 r^{\alpha_1} r^{\alpha_2} r^{\alpha_3} r^{\alpha_4} + B_3 \delta^{\alpha_1\alpha_3} \delta^{\alpha_2\alpha_4} + B_4 r^{\alpha_1} r^{\alpha_3} \delta^{\alpha_2\alpha_4}, \qquad (2.3.4)$$

with the coefficients $B_0,\ldots,B_5$ for all types of lateral interactions of nonpolar molecules listed in Table 2.2.

Table 2.2. Determination of interaction constants

Quantity	Quadrupole	Dispersion	Repulsion
B_0	(3/4)U	$-2W_2 - 4W_3$	$(1-12\alpha^2+84\alpha^4)R$
B_1	-(15/4)U	$-3(W_2 - W_3)$	$84\alpha^2(1-16\alpha^2)R$
B_2	(105/4)U	$-9(W_1 - 2W_2 + W_3)$	$12096\alpha^4 R$
B_3	(3/2)U	$-(W_1 - 2W_2 + W_3)$	$84\alpha^4 R$
B_4	-15U	$6(W_1 - 2W_2 + W_3)$	$-2688\alpha^4 R$
B_5	0	0	$2016\alpha^4 R$

Characteristic energies of the interactions concerned are defined by the following relations:

$$U=\frac{Q^2}{a^5},\quad R=\frac{4K}{a^{12}},\quad \alpha=\frac{d}{a},\quad W_1=\frac{\hbar}{2\pi a^6}\int_0^\infty \chi_\parallel^2(i\omega)d\omega,$$
$$W_2=\frac{\hbar}{2\pi a^6}\int_0^\infty \chi_\parallel(i\omega)\chi_\perp(i\omega)d\omega,\quad W_3=\frac{\hbar}{2\pi a^6}\int_0^\infty \chi_\perp^2(i\omega)d\omega, \tag{2.3.5}$$

where Q is the molecular quadrupole moment, $a = |\mathbf{r}_{jj'}|$ is the intermolecular distance, K is the repulsive potential parameter, $2d$ is the length of the molecular bond, $\chi_\parallel(\omega)$ and $\chi_\perp(\omega)$ designate polarizability components longitudinal and transverse with respect to the long molecular axis.

It is noteworthy that some terms originating from lateral interactions are once summed over molecular numbers j and merely renormalize, in this sense, the adsorption potential $U(\theta_j,\varphi_j)$ for each molecule. The other terms to be summed over j and j' represent cooperative effects and are sensitive solely to coefficients B_2, B_3, and B_4 which include only small-magnitude anisotropic combinations of dispersion and repulsion interactions,

$$W_1-2W_2+W_3=\frac{\hbar}{2\pi a^6}\int_0^\infty [\chi_\parallel(i\omega)-\chi_\perp(i\omega)]^2 d\omega \tag{2.3.6}$$

and $\alpha^4 R$, which are in addition have opposite sign. These small and mutually compensated contributions result in a negligible additional renormalization of the contribution from quadrupole-quadrupole interactions. The Table 2.3 presents the contributions in question estimated for the system CO/NaCl(100) with the parameters Q=1.62×10^{-26} esu cm^2 and a=5.64/√2 Å, and literature values of other parameters.[47,49]

Table 2.3. Estimated interaction constants for CO/NaCl(100) (meV)

U	$W_1 - 2W_2 + W_3$	$84\alpha^4 R$	μ^2/a^3
1.63	0.5	0.3	0.1

As mentioned above, the energy of dipole-dipole interactions in this system also proves to be small enough.

A specific role of adsorption potentials $U(\theta_j,\varphi_j)$ should be pointed out here. In addition to the fact that they are characterized by angular dependences, they also fix molecular centers of mass at certain points on the surface thus affording the orientational and translational aspects of the problem to be treated separately. If such potentials were absent, equilibrium intermolecular distances a would result from all sorts of competing interactions; it would therefore be impossible to ignore dispersion and repulsion forces, if for no other reason than their contribution to the angle-average interaction energy for each pair of molecules:

$$\left\langle H_{\text{or}}^{(jj')} \right\rangle\Big|_{U=0} = -\frac{2}{3}(W_1 + 4W_2 + 4W_3) + \left(1 + 44\alpha^2 + \frac{16016}{15}\alpha^4\right)R. \qquad (2.3.7)$$

The above equation following from relations (2.3.1)-(2.3.4) reduces, with the values listed in Table 2.2, to known expressions for dispersion[80] and repulsion[78] interactions and proves helpful in the experiment-based estimation of the parameters entering into it.

For diatomic molecules adsorbed by metal surfaces, the adsorption potentials $U(\theta_j, \varphi_j)$ fix, as a rule, molecular axis perpendicular to the surface (θ_j=0) thus cancelling the question about optimum orientational structures. For dielectric substrates, inclined molecular orientations emerge, with their specificity governed by the competition of adsorption potentials and lateral interactions. Experimental evidence for the system CO/NaCl(100)[28] demonstrates that inclination angles θ_j prove to be the same for all the molecules whereas azimuthal angles φ_j alternate so that the sublattices of identically oriented molecules arise. The constant value of the angle θ_j is attributable both to a possible deep minimum of the function $U(\theta_j, \varphi_j)$ alone and to the corresponding minima of the whole function $H_{\text{or}}(\theta_j)$ which may result from competing adsorption potentials and lateral interactions. The approximation of equal angles θ_j is quite reasonable in the former case and needs additional verification in the latter, with its validity easily verified by the analysis of temperature dependences of inclination angles.

A theoretical treatment of the effect caused by the competition between the sine-like angular-dependent component of the adsorption potential and dipole lateral interaction demonstrated that the values θ_j are the same in the ground state and at the phase transition temperature.[81] Study of the structure and dynamics for the CO monolayer adsorbed on the NaCl(100) surface using the molecular dynamics method has also led to the inference that angles θ_j are practically equalized in a wide temperature range.[82] That is why the following consideration of orientational structures and excitations in a system of adsorbed molecules will imply, for the sake of simplicity, the constant value of the inclination angle $\theta_j = \theta$ (see Fig. 2.14) which is due to the adsorption potential $U(\theta_j, \varphi_j)$.

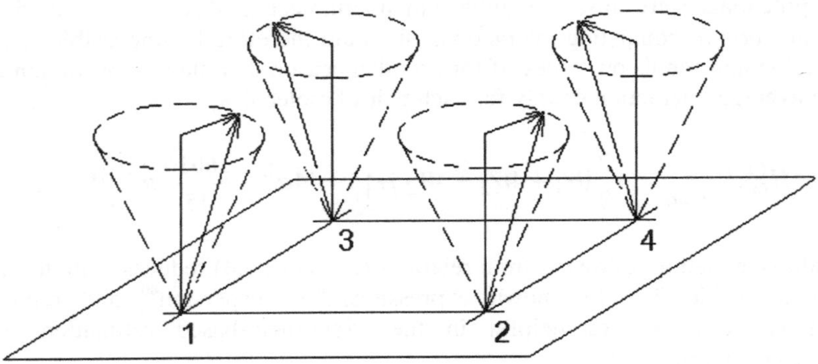

Fig. 2.14. A schematic representation of orientational structure for adsorbed molecules inclined at the same angle to the surface normal.

Introducing two-dimensional unit vectors ξ_j for azimuthal orientations of adsorbed molecules

$$\mathbf{e}_j = \mathbf{n}\cos\theta + \xi_j \sin\theta\ , \qquad \xi_j = (\cos\varphi_j, \sin\varphi_j, 0) \tag{2.3.8}$$

and substituting them into Eq. (2.3.1) we arrive at:

$$H_{\text{or}} = \sum_j \left[U(\varphi_j) + \xi_j^{\beta_1}\xi_j^{\beta_2}\xi_j^{\beta_3}\xi_j^{\beta_4} \sum_{j'(\neq j)} \widetilde{V}'^{\beta_1\beta_2\beta_3\beta_4}(\mathbf{r}_{jj'}) \right]$$
$$+\frac{1}{2}\sum_{j\neq j'}\left[V^{\beta_1\beta_2}(\mathbf{r}_{jj'})\cos^2\theta\sin^2\theta\,\xi_j^{\beta_1}\xi_{j'}^{\beta_2} + V^{\beta_1\beta_2\beta_3\beta_4}(\mathbf{r}_{jj'})\sin^4\theta\,\xi_j^{\beta_1}\xi_j^{\beta_2}\xi_{j'}^{\beta_3}\xi_{j'}^{\beta_4}\right], \tag{2.3.9}$$

where twice-used indices $\beta_1,\ldots,\beta_4$ imply summation only over the x and y components of the corresponding vectors and tensors, and two new tensors are introduced:

$$\widetilde{V}'^{\beta_1\beta_2\beta_3\beta_4}(\mathbf{r}) = \frac{1}{2}\left(B_0 + B_3\cos^4\theta\right)\delta^{\beta_1\beta_2}\delta^{\beta_3\beta_4} + B_1\sin^2\theta\, r^{\beta_1} r^{\beta_2}\delta^{\beta_3\beta_4}$$
$$+ B_5\sin^4\theta\, r^{\beta_1} r^{\beta_2} r^{\beta_3} r^{\beta_4}\,, \tag{2.3.10}$$

$$V^{\beta_1\beta_2}(\mathbf{r}) = 2B_3\delta^{\beta_1\beta_2} + B_4 r^{\beta_1} r^{\beta_2}\,. \tag{2.3.11}$$

The function $U(\varphi_j)$ describes the azimuthal angular dependence of the adsorption potential $U(\theta_j, \varphi_j)$ with the θ_j fixed.

An inference of fundamental importance follows from Eqs. (2.3.9) and (2.3.11): When long axes of nonpolar molecules deviate from the surface-normal direction slightly enough, their azimuthal orientational behavior is accounted for by much the same Hamiltonian as that for a two-dimensional dipole system. Indeed, at $\sin\theta << 1$ the main nonlocal contribution to Eq. (2.3.9) is provided by a term quadratic in ξ_j which contains the interaction tensor $V^{\beta_1\beta_2}(\mathbf{r})$ of much the same structure as dipole-dipole interaction tensor: $2B_3 > 0$, $B_4 < 0$, only differing in values $2B_3$ and B_4. For dipole-dipole interactions, $2B_3 = D = \mu^2/a^3$ (μ is the dipole moment) and $B_4 = -3D$, whereas, e.g., purely quadrupole-quadrupole interactions are characterized by $2B_3 = 3U$, $B_4 = -15U$ (see Table 2.2). Evidently, it is for this reason that the dipole model applied to the system CO/NaCl(100), with rather small values θ ($\theta \approx 25^0$), provided an adequate picture for the ground-state orientational structure.[81] A contradiction arose only in the estimation of the temperature T_c of the observed orientational phase transition: For the experimental value T_c = 25 K to be reproduced, the dipole moment should have been set μ = 1.3D, which is ten times as large as the corresponding value μ in a gas phase. Section 2.4 will be devoted to a detailed consideration of orientational states and excitation spectra of a model system on a square lattice described by relations (2.3.9)-(2.3.11).

There is another case when the orientational Hamiltonian for nonpolar molecules (2.3.1) on a symmetric two-dimensional lattice is reducible to a quasidipole form, viz. the case of planar orientations of long molecular axes ($\theta_j = 90^0$ in expression (2.3.2)) when one can invoke the transformation for doubled orientation angles φ_j of the unit vectors $\mathbf{e}_j$ and $\mathbf{r}_{jj'}$:

$$\mathbf{e}_j \otimes \mathbf{e}_j = \mathbf{S}_0 + \mathbf{S}_\gamma\,\varepsilon_j^\gamma\,, \quad \boldsymbol{\varepsilon}_j = \left(\cos 2\varphi_j, \sin 2\varphi_j, 0\right),$$
$$\mathbf{r}_{jj'} \otimes \mathbf{r}_{jj'} = \mathbf{S}_0 + \mathbf{S}_\gamma\,\rho_{jj'}^\gamma\,. \tag{2.3.12}$$

(Here the symbol $\otimes$ denotes the direct vector product and the matrices $\mathbf{S}_0$ and $\mathbf{S}_\gamma$ ($\gamma = x, y$) resemble Pauli matrices (accurate to factors and redesignated indices))

$$\mathbf{S}_0 = \frac{1}{2}\begin{pmatrix} 1 & 0 \\ 0 & 1 \end{pmatrix}, \; \mathbf{S}_x = \frac{1}{2}\begin{pmatrix} 1 & 0 \\ 0 & -1 \end{pmatrix}, \; \mathbf{S}_y = \frac{1}{2}\begin{pmatrix} 0 & 1 \\ 1 & 0 \end{pmatrix} \qquad (2.3.13)$$

which satisfy the identities:

$$\mathrm{Sp}(\mathbf{S}_0 \cdot \mathbf{S}_0) = \frac{1}{2}, \; \mathrm{Sp}(\mathbf{S}_\gamma \cdot \mathbf{S}_{\gamma'}) = \frac{1}{2}\delta_{\gamma\gamma'}, \; \mathrm{Sp}(\mathbf{S}_\gamma \cdot \mathbf{S}_0) = 0 . \qquad (2.3.14)$$

The above expressions enable the vectors $\boldsymbol{\varepsilon}_j$ and $\boldsymbol{\rho}_{jj'}$ of the doubled angles to be expressed in terms of the vectors $\mathbf{e}_j$ and $\mathbf{r}_{jj'}$ of the initial angle values:

$$\varepsilon_j^\gamma = 2\,\mathrm{Sp}(\mathbf{S}_\gamma \cdot \mathbf{e}_j \otimes \mathbf{e}_j) \;,\; \rho^\gamma = 2\,\mathrm{Sp}(\mathbf{S}_\gamma \cdot \mathbf{r} \otimes \mathbf{r}) \; . \qquad (2.3.15)$$

As a result, intermolecular potential (2.1.2) assumes the form:

$$\begin{aligned} V(\mathbf{r}, \mathbf{e}_j, \mathbf{e}_{j'}) &= \frac{1}{4}(4B_0 + 4B_1 + B_2 + 2B_3 + B_4 + 2B_5) \\ &+ \frac{1}{4}(2B_1 + B_2 + B_4 + 2B_5)\boldsymbol{\rho} \cdot (\boldsymbol{\varepsilon}_j + \boldsymbol{\varepsilon}_{j'}) + \frac{1}{4}B_5\left[(\boldsymbol{\rho} \cdot \boldsymbol{\varepsilon}_j)^2 + (\boldsymbol{\rho} \cdot \boldsymbol{\varepsilon}_{j'})^2\right] \\ &+ \frac{1}{4}(2B_3 + B_4)(\boldsymbol{\varepsilon}_j \cdot \boldsymbol{\varepsilon}_{j'}) + \frac{1}{4}B_2(\boldsymbol{\rho} \cdot \boldsymbol{\varepsilon}_j)(\boldsymbol{\rho} \cdot \boldsymbol{\varepsilon}_{j'}) \, . \end{aligned} \qquad (2.3.16)$$

On summing this expression over all pairs of neighboring molecules on symmetric two-dimensional lattices, the second term gives no contribution to the sum, and the third one causes only a negligible correction. The main contributions are provided by the last two terms which together make up the structure of the dipole Hamiltonian, with the constants determined by the parameters of the interactions considered.

Substituting Eq. (2.3.16) into Eq. (2.3.9) (with $\theta = 90^0$), we arrive at the following expression referring to square and triangular lattices:

$$H_{\mathrm{or}} = \sum_j \left[U(\varphi_j) + V(\boldsymbol{\varepsilon}_j)\right] + \frac{1}{2}\sum_{j \neq j'} V^{\gamma_1\gamma_2}(\boldsymbol{\rho}_{jj'})\varepsilon_j^{\gamma_1}\varepsilon_{j'}^{\gamma_2} \, , \qquad (2.3.17)$$

where

$$V(\varepsilon_j)=\begin{cases}\frac{1}{2}(4B_0+4B_1+B_2+2B_3+B_4+2B_5)+B_5\left(\varepsilon_j^x\right)^2 & \text{– square lattice}\\ \frac{3}{4}(4B_0+4B_1+B_2+2B_3+B_4+3B_5) & \text{– triangular lattice}\end{cases}, \quad (2.3.18)$$

and the tensor

$$V^{\gamma_1\gamma_2}(\mathbf{\rho})=D_1\delta^{\gamma_1\gamma_2}+D_2\rho^{\gamma_1}\rho^{\gamma_2} \quad (2.3.19)$$

again takes on a quasidipole form with the constants

$$D_1=\frac{1}{4}(2B_3+B_4),\quad D_2=\frac{1}{4}B_2. \quad (2.3.20)$$

Note that the dependence of lateral interactions defined by Eq. (2.3.19) on the azimuthal angles φ_i, φ_j, θ_{ij} corresponding to the vectors $\hat{\mathbf{e}}_i$, $\hat{\mathbf{e}}_j$, $\hat{\mathbf{r}}_{ij}$ is expressed explicitly as

$$\begin{aligned}V^{\gamma_1\gamma_2}(\mathbf{\rho}_{jj'})\varepsilon_j^{\gamma_1}\varepsilon_{j'}^{\gamma_2}&=\left(D_1+\frac{1}{2}D_2\right)\cos(2\varphi_j-2\varphi_{j'})\\&+\frac{1}{2}D_2\cos(2\varphi_j+2\varphi_{j'}-4\theta_{jj'})\end{aligned} \quad (2.3.21)$$

being thus specified just by the doubled angles. A representation of this kind was applied formerly [83,84] to treat a planar system of quadrupoles on a triangular lattice. For quadrupole interactions, $2D_1 + D_2 \ll D_2$ and only the second term of relation (2.3.21) was taken into account in Refs. 83,84.

It should be noted that the first sum in expression (2.3.17) describes the local contributions to the orientational Hamiltonian from the adsorption potentials $U(\varphi_j)$ and from the intermolecular interactions $V(\varepsilon_j)$. The latter term is angular-dependent only for a square lattice, which is due to repulsive interactions; this dependence is slight owing to the coefficient B_5 smallness (see Eq. (2.3.18) and Table 2.2). As $B_5 > 0$, the term $V(\varepsilon_j)$ for a square lattice causes the values ε_j^x to decrease at short intermolecular distances and the vectors $\mathbf{e}_j$ therefore have a feeble tendency to orient along square lattice diagonals. The angular dependence of the potentials $U(\varphi_j)$ accounts for the symmetry of the substrate atoms arrangement and is characterized by several equivalent minima. That is why the orientational structure of adsorbed molecules lying in the surface plane is dictated by the nonlocal interaction component, i.e., by the second sum in the Eq. (2.3.17). (A specific case of the competition between the adsorption potential $U(\varphi_j)$ and local interactions in forming preferable orientations is pursued further in Sec. 2.5). This sum has the same form as Eq. (2.2.1) and the analysis of its minimum values is similar to the analysis presented in Sec. 2.2 differing, however, in two points. First, the vectors $\mathbf{r}_{jj'}$ and $\mathbf{e}_j$ are substituted by the corresponding vectors of the doubled angles, $\boldsymbol{\rho}_{jj'}$ and $\boldsymbol{\varepsilon}_j$ and, second, the coefficients D_1 and D_2 defined by Eq. (14) can take on arbitrary values due to the competition of quadrupole, dispersion, and repulsive interactions. The minimization procedure is additionally simplified, as the Fourier components of the tensor (2.3.19)

$$\widetilde{V}^{\gamma_1\gamma_2}(\mathbf{q}) = \sum_{j'} V^{\gamma_1\gamma_2}(\boldsymbol{\rho}_{jj'})\exp(-i\mathbf{q}\cdot\mathbf{r}_{jj'}) \qquad (2.3.22)$$

can be represented, by virtue of the short-range interaction nature, in a simple analytical form. For square and triangular lattices, the following expressions are derived:

$$\widetilde{V}^{\gamma_1\gamma_2}(\mathbf{q}) = 2\begin{pmatrix} D_1 + D_2 & 0 \\ 0 & D_1 \end{pmatrix}(\cos q_x + \cos q_y), \qquad (2.3.23)$$

$$\begin{aligned}\widetilde{V}^{\gamma_1\gamma_2}(\mathbf{q}) &= 2D_1\begin{pmatrix} 1 & 0 \\ 0 & 1 \end{pmatrix}\left(\cos q_x + 2\cos(q_x/2)\cos\left(\sqrt{3}\, q_y/2\right)\right) \\ &+ D_2\begin{pmatrix} 2\cos q_x + \cos(q_x/2)\cos\left(\sqrt{3}\, q_y/2\right) & \sqrt{3}\sin(q_x/2)\sin\left(\sqrt{3}\, q_y/2\right) \\ \sqrt{3}\sin(q_x/2)\sin\left(\sqrt{3}\, q_y/2\right) & 3\cos(q_x/2)\cos\left(\sqrt{3}\, q_y/2\right) \end{pmatrix},\end{aligned} \qquad (2.3.24)$$

It is easily seen that the minimum eigenvalues for tensors (2.3.21) and (2.3.22), corresponding to the unit lengths of the vectors $\boldsymbol{\varepsilon}_j$, can result only at the symmetric points of the first Brillouin zone, namely, at $\mathbf{q}_0 = (0, 0, 0)$, $\mathbf{q}_J = (\pi, \pi, 0)$ for a square lattice and at $\mathbf{q}_0 = (0, 0, 0)$, $\mathbf{q}_A = (0, 2\pi/\sqrt{3}, 0)$, $\mathbf{q}_J = (4\pi/3, 0, 0)$ for a triangular lattice. Due to the isotropy of tensor (2.3.22) at the points $\mathbf{q}_0$ and $\mathbf{q}_J$, there exist only four variants of eigenvalues both for square and triangular lattices. For a square lattice whose lattice units are specified by the radius-vectors $\mathbf{r} = (m, n, 0)$ (m and n are integers), these eigenvalues and the corresponding azimuthal angles of orientational structures, φ_{mn}, appear as:

$$\begin{aligned}
&\tilde{V}^{xx}(\mathbf{q}_0) = 4(D_1 + D_2), \quad \varphi_{mn} = 0, \\
&\tilde{V}^{yy}(\mathbf{q}_0) = 4D_1, \quad \varphi_{mn} = \pi/4, \\
&\tilde{V}^{xx}(\mathbf{q}_J) = -4(D_1 + D_2), \quad \varphi_{mn} = \pi/4\left[1 - (-1)^{m+n}\right], \\
&\tilde{V}^{yy}(\mathbf{q}_J) = -4D_1, \quad \varphi_{mn} = \pi/4\left[2 - (-1)^{m+n}\right].
\end{aligned} \tag{2.3.25}$$

The listed values $\tilde{V}$, if multiplied by $N/2$ (half the number of molecules) provide the energies of the nonlocal interaction components. To find the ground states, the values $\tilde{V}$ should be compared in magnitude. Designate the orientational phases (2.3.25) by $\mathbf{q}_0\backslash X$, $\mathbf{q}_0\backslash Y$, $\mathbf{q}_J\backslash X$, and $\mathbf{q}_J\backslash Y$. The corresponding interphase boundaries will then be described as follows:

$$\begin{aligned}
&\mathbf{q}_0 \backslash Y \leftrightarrow \mathbf{q}_0 \backslash X: \quad D_2 = 0, \ D_1 < 0, \\
&\mathbf{q}_0 \backslash X \leftrightarrow \mathbf{q}_J \backslash Y: \quad D_2 = -2D_1, \ D_1 > 0, \\
&\mathbf{q}_J \backslash Y \leftrightarrow \mathbf{q}_J \backslash X: \quad D_2 = 0, \ D_1 > 0, \\
&\mathbf{q}_J \backslash X \leftrightarrow \mathbf{q}_0 \backslash Y: \quad D_2 = -2D_1, \ D_1 < 0.
\end{aligned} \tag{2.3.26}$$

Thus, we are led to the phase diagram shown in Fig. 2.15.

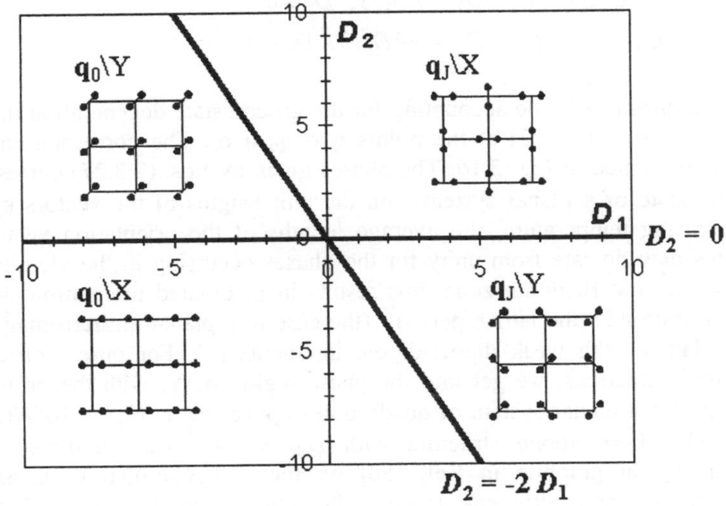

Fig. 2.15. Planar orientational structures of nonpolar molecules on a square lattice.

The ground states of the various nature arise depending on the parameter values D_1 and D_2. A phase diagram of this kind which is concerned with the interactions of diatomic molecules via atom-atom Lennard-Jones potentials and makes allowance for the angular dependence of the local part of C_4-symmetric interactions (see Eq. (2.3.18) for a square lattice) was constructed previously.[85] As an example, for purely quadrupole-quadrupole interactions we have $D_1 = -3U$, $D_2 = 105U/16$, which is typical of the phase region $\mathbf{q}_J\backslash X$ with T-like molecular orientations. The total energy of this ground state (with the local contribution (2.3.18) included) is found to be $H_{or} = -6NU$.

For a triangular lattice with the lattice units $\mathbf{r} = (m+n/2, \sqrt{3}n/2, 0)$, we derive likewise:

$$
\begin{aligned}
&\tilde{V}^{xx}(\mathbf{q}_0) = \tilde{V}^{yy}(\mathbf{q}_0) = 6D_1 + 3D_2, \quad \varphi_{mn} = \alpha, \\
&\tilde{V}^{xx}(\mathbf{q}_J) = \tilde{V}^{yy}(\mathbf{q}_J) = -3D_1 - 3D_2/2, \quad \varphi_{mn} = \alpha + 2\pi m/3 + \pi n/3, \\
&\tilde{V}^{xx}(\mathbf{q}_A) = -2D_1 + D_2, \quad \varphi_{mn} = \pi/4\left[1 - (-1)^n\right], \\
&\tilde{V}^{yy}(\mathbf{q}_A) = -2D_1 - 3D_2, \quad \varphi_{mn} = \pi/4\left[2 + (-1)^n\right],
\end{aligned}
\tag{2.3.27}
$$

$$
\begin{aligned}
&\mathbf{q}_0 \leftrightarrow \mathbf{q}_A \backslash X: \quad D_2 = -4D_1, \; D_1 > 0, \\
&\mathbf{q}_A \backslash X \leftrightarrow \mathbf{q}_J: \quad D_2 = -2D_1/5, \; D_1 > 0, \\
&\mathbf{q}_J \leftrightarrow \mathbf{q}_A \backslash Y: \quad D_2 = 2D_1/3, \; D_1 > 0, \\
&\mathbf{q}_A \backslash Y \leftrightarrow \mathbf{q}_0: \quad D_2 = -4D_1/3, \; D_1 < 0.
\end{aligned}
\tag{2.3.28}
$$

Here α is an arbitrary angle accounting for the ground state degeneration due to the isotropy of tensors (2.3.22) at the points $\mathbf{q} = \mathbf{q}_0$ и $\mathbf{q}_J$. The corresponding phase diagram is presented in Fig. 2.16. The phases given by Eqs. (2.3.27) correspond to the ground state of a planar system with the unit lengths of the vectors ε_j and $\mathbf{e}_j$. With increasing temperature, the average lengths of the orientation vectors $\mathbf{e}_j$ at lattice sites may deviate from unity for the phases occurring in the vicinity of the point J of the first Brillouin zone; this results in modulated orientation structures incommensurate with the lattice period[86] (the case of a planar antiferromagnet on a triangular lattice with weak dipole-dipole interactions[87]). For purely quadrupole-quadrupole interactions, we get into the phase region $\mathbf{q}_A\backslash Y$, with the ground-state total energy for a planar system of quadrupoles appearing as $H_{or} = -165NU/32 \approx -5.16NU$. This herringbone structure with $\varphi_{mn} \approx \pm 45^0$ was predicted for the monolayer N_2 on graphite in Ref. [88] on the computational basis assuming graphite-nitrogen and nitrogen-nitrogen interaction potential of Lennard-Jones form, and quadrupole-quadrupole interactions between nearest-neighbor nitrogen molecules. The two-sublattice structure given by $\varphi_{mn} = \pm 45^0$ is identical to the two-sublattice in-plane phase predicted by Harris and Berlinsky[89] for large crystal fields and low temperatures.

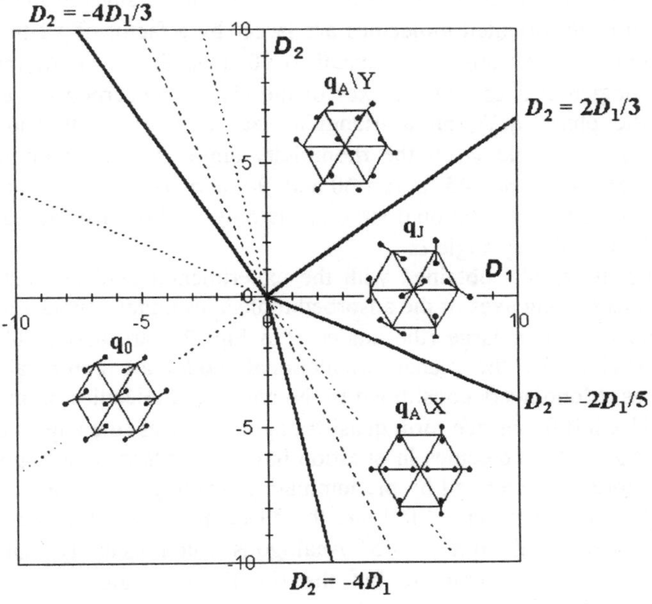

Fig. 2.16. Planar orientational structures of nonpolar molecules on a triangular lattice.

Configurations closely resembling this one were also observed in the Monte Carlo simulations and regarded to be energetically optimum for 36 quadrupoles constrained to a plane in a triangular lattice.[90] Improved Monte Carlo simulations for a triangular lattice with 400, 1600, 6400, and 10000 sites were performed without[83] and with[91] regard to quantum effects. The dynamical excitations in the system N_2/graphite were also studied using molecular-dynamics simulations.[92] The ground state for three-dimensional orientations of quadrupoles on a triangular lattice is characterized by the four-sublattice orientational structure $\theta_1 = 0$; $\theta_2 = \pi/2$, $\varphi_2 = \alpha$; $\theta_3 = \pi/2$, $\varphi_3 = \alpha + \pi/3$; $\theta_4 = \pi/2$, $\varphi_4 = \alpha + 2\pi/3$ and a somewhat smaller energy $H_{or} = -765NU/128 \approx -5.98NU$; this ground state grades into a planar structure $\mathbf{q}_A\backslash Y$ under the action of the adsorption potential tending to orient molecules parallel to the surface plane.[89]

It should be pointed out that quadrupole-quadrupole molecular interactions on square and triangular lattices cause the same orientational structures. Indeed, the phase $\mathbf{q}_J\backslash X$ on a square lattice can be regarded as alternating chains of identically oriented molecules, with the chain axes parallel to one of the diagonals of the unit-cell square. Thus, the structure arises in which the square sides in the sublattices a and b of identically oriented molecules are larger by a factor of $\sqrt{2}$ than the initial lattice constant, and the angle α is equal to 90^0 (see Fig. 2.4). Azimuthal angles φ between molecular axes and the axes of the chains concerned are equal to 45^0, just as in the phase $\mathbf{q}_A\backslash Y$ on a triangular lattice. It is easily shown that the dependence of the angle φ on the rhombicity angle α is nonmonotonic on the interval from 60 to 90^0: $\varphi = 45^0$ at $\alpha = 60$ and 90^0 and $\varphi \approx 43^0$ (the minimum value) at $\alpha \approx 72^0$. Thus, the orientational structure is only slightly distorted at significant changes in the rhombicity angle α.

Correlate the results obtained with the experimental evidence listed in Table 2.1. Projections of long axes of the adsorbed molecules under consideration onto the surface plane are rather large (the angles θ in Fig. 2.4 are about 90^0 so that the structures provided by the planar orientational model are comparable with the observed ones. Monolayers constituted by N_2 and CO_2 molecules on graphite or by CO_2 on the NaCl(100) surface form quasi-triangular lattices (the angles α in Fig. 2.4 are close to 60^0). The molecules in question have rather large quadrupole moments and are therefore characterized by predominating quadrupole interactions; thus, the orientational structures observed fall into the phase $\mathbf{q}_A\backslash Y$, with the azimuthal angles φ close to the theoretical value $\varphi = 45^0$. Analogous structures are typical of the C_2N_2 and CS_2 monolayers on graphite but the rhombicity angle α of their lattices markedly differs from 60^0. On the other hand, carbon atoms of the substrate form a hexagonal (honeycomb) lattice, with its symmetry different from the symmetry of the adsorbate lattice. As a consequence, the local contributions from adsorption potentials $U(\varphi_j)$ to the orientational Hamiltonian can cause sufficiently large deviations of the equilibrium angles φ from 45^0. Notice that in the case of the C_2N_2 and CS_2 monolayers, the φ deviation of 10^0 from the value 45^0 cannot be accounted for by the changes in the nonlocal interaction component caused by the rhombic lattice distortion, as this effect can only result in φ deviations of no more than 2^0.

To treat the orientational structure of the monolayer formed by O_2 molecules on a graphite surface, allowance must be made for the fact that an oxygen molecule is characterized not only by a nonzero magnetic moment but also by a record small quadrupole moment, so that dispersion interactions prevail over quadrupole interactions at intermolecular distances shorter than 10 Å.[79] In addition, the adsorbate lattice parameters give rise to very small minimum intermolecular distances, $a \approx 3.3$ Å, the parameter $b \approx 8.1$ Å markedly exceeding the values a. That is why, it is sufficient to consider only the nearest-neighbor interactions in a molecular chain, i.e., between the molecules spaced at the distance a. The latter problem is reduced to the equilibrium orientation determination for a corresponding molecular dimer. For the O_2 dimer, the change in a from 3.1 до 4.2 Å results in the energetic preference of H-configurations in which long molecular axes are perpendicular to the line connecting the centers of molecules.[93] Thus, the H-configurations with $\varphi = 90^0$ are realized in the O_2 monolayer on a graphite surface.

Interesting comparisons can be made between other orientational structures listed in Table 2.1 and the experimental evidence on corresponding dimers. T-configurations, with one molecule parallel and the other perpendicular to the line connecting their centers, are known to be energetically advantageous when quadrupole interactions predominate. To exemplify, the T-configuration for the N_2 dimer results at $a > 4.1$ Å.[94,95] Since quadrupole moments of the CO_2, C_2N_2, and CS_2 molecules are not smaller than that of the N_2, they also should exhibit T-configurations at the distances a indicated in Table 2.1. For a square lattice, it turns out that the molecules can orient so that T-configurations arise in each pair of neighboring lattice sites. The resulting structure just falls into the above-considered $\mathbf{q}_J$\X phase. In the case of rhombic lattices with $\alpha \neq 90^0$, T-configurations cannot involve all the neighboring lattice sites; the minimization of the full Hamiltonian (2.3.1) is therefore necessary, with due regard not only for quadrupole interactions but for other interaction types of nonpolar molecules as well.

2.4. Ground states and phase transitions for a system of nonpolar molecules on a square lattice

Ground-state structures dictated by the Hamiltonian (2.3.9) are first of all dependent on the form of the function $U(\varphi_j)$. For instance, it is reasonable to assume for a square lattice that the function $U(\varphi_j)$ has a fourth-order symmetry and can be approximated by the hindered-rotation potential

$$U(\varphi_j) = \frac{1}{2}\Delta U_4 \cos 4\varphi_j \qquad (2.4.1)$$

in which positive values of the reorientation barrier ΔU_4 specify such equilibrium orientations for projections of long molecular axes which are parallel to square lattice diagonals (φ_j = 45, 135, 225, 315^0), whereas negative values correspond to orientations parallel to lattice axes (φ_j = 0, 90, 180, 270^0). Then the Hamiltonian defined by Eq. (2.3.9), with relations (2.3.10) and (2.3.11) included, takes the form:

$$H_{\text{or}} = 2N\left(B_0 + B_1 \sin^2\theta + B_3\cos^4\theta - \frac{3}{4}\Delta U_4 \right)$$
$$+\sum_{nm}\Big\{\Big[2B_3\left(\xi_{nm}\cdot\xi_{n,m+1}+\xi_{nm}\cdot\xi_{n+1,m}\right)+B_4\left(\xi^x_{nm}\xi^x_{n,m+1}+\xi^y_{nm}\xi^y_{n+1,m}\right)\Big]\cos^2\theta\sin^2\theta$$
$$+\Big[B_2\Big(\left(\xi^x_{nm}\right)^2\left(\xi^x_{n,m+1}\right)^2+\left(\xi^y_{nm}\right)^2\left(\xi^y_{n+1,m}\right)^2\Big)+B_3\Big(\left(\xi_{nm}\cdot\xi_{n,m+1}\right)^2+\left(\xi_{nm}\cdot\xi_{n+1,m}\right)^2\Big)$$
$$+B_4\Big(\xi^x_{nm}\xi^x_{n,m+1}\left(\xi_{nm}\cdot\xi_{n,m+1}\right)+\xi^y_{nm}\xi^y_{n+1,m}\left(\xi_{nm}\cdot\xi_{n+1,m}\right)\Big)\Big]\sin^4\theta$$
$$+2\left(B_5\sin^4\theta+\Delta U_4\right)\Big(\left(\xi^x_{nm}\right)^4+\left(\xi^y_{nm}\right)^4\Big)\Big\}, \tag{2.4.2}$$

where the sites $\mathbf{r}_j = (m, n, 0)$ of a square lattice are given by a couple of indices, n and m.

In most cases, ground states of lattice systems can be restricted to structures with a doubled lattice period.[55,96-98] Within this structure set, the minimum values of expression (2.4.2) at $\Delta U_4 = 0$ correspond to various orientational phases, depending on the angle θ between molecular axes and the surface-normal directions. For purely quadrupole-quadrupole interactions, characteristics of the phases are presented in Table 2.4 and in Fig. 2.17 (allowance made for dispersion and repulsion forces would result only in a slight renormalization of the parameter values). As mentioned in Section 2.3, the main contribution to the energy of the system at small θ arises from dipole-like interactions. The minimum values of the corresponding terms at $B_4<-2B_3<0$ are equal to NB_4 and specify a 2x2 orientational structure which is continuously degenerate in the azimuthal inclination angle φ reckoned from square lattice axes and obeying the relations $\varphi \equiv \varphi_1 = -\varphi_2 = 180^0-\varphi_3 = \varphi_4 - 180^0$. The other terms in Eq. (2.4.2) remove the degeneration selecting the structure with $\varphi = 45^0$ (the portion AC in Fig. 2.17) as the most advantageous one. However, another structure, with $\varphi = 0$ (the portion AB in Fig. 2.17), which is characterized by the gain in free energy of order[70]

$$\Delta F \approx -0.1Nk_{\text{B}}T\left(\frac{B_4+2B_3}{B_4-2B_3}\cos 2\varphi\right)^2, \tag{2.4.3}$$

becomes preferential as the temperature increases (due to thermodynamic fluctuations).

Table 2.4. Characteristics of the orientational phases

Phase, Phase Point	θ (degree)	φ_j (degree)	H_{or}/NU ($t = \sin^2\theta$)
A-B A-E	[0, 24.6] [0, 45.7]	$\varphi_1 = \varphi_2 = 0$ $\varphi_3 = \varphi_4 = 180$	$(9/2)-(57/2)t+(129/4)t^2$
E-G	[45.7, 90]	$\varphi_1 = \varphi_4 = 0$ $\varphi_2 = \varphi_3 = 90$	$(9/2)-(27/2)t+3t^2$
A-C	[0, 30.6]	$\varphi_1 = -\varphi_2 = 45$ $\varphi_3 = -\varphi_4 = 135$	$(9/2)-(57/2)t+(249/8)t^2$
C	30.6	$\varphi_1 = \varphi_2 = 26.5$ $\varphi_3 = \varphi_4 = 153.5$	($t = 0.2589$)
B(C)-D	[24.6 (30.6), 37.5]	$\varphi_1 = \varphi_2$ $=(1/2)\arccos[4(1-t)/19t]$ $\varphi_3 = \varphi_4 = 180 - \varphi_1$	$(159/38)-(945/38)t+(3315/152)t^2$
D	37.5	$\varphi_1 = \varphi_2 = 34.5$ $\varphi_3 = \varphi_4 = 145.5$	($t = 0.3708$)
D-F	[37.5, 63.4]	$\varphi_1 = \varphi_2 = \varphi_3 = \varphi_4 = 45$	$(9/2)-(45/2)t+(105/8)t^2$
F-G	[63.4, 90]	$\varphi_1 = \varphi_4 =$ $(1/2)\arcsin[4(1-t)/t]$ $\varphi_2 = \varphi_3 = 90 - \varphi_1$	$-(27/2)+(45/2)t-15t^2$

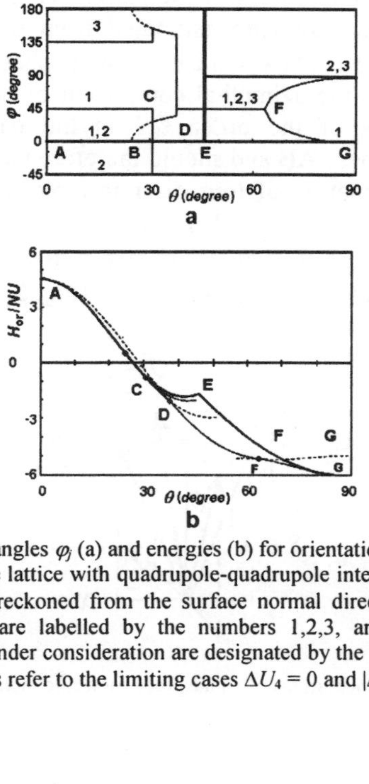

Fig. 2.17. Azimuthal angles φ_j (a) and energies (b) for orientational structures of adsorbed molecules on a square lattice with quadrupole-quadrupole interactions plotted versus the inclination angle θ reckoned from the surface normal direction. Sublattices j of the structures described are labelled by the numbers 1,2,3, and the boundaries of the orientational phases under consideration are designated by the letters A, B, C, D, E, F, G. Regular and bold lines refer to the limiting cases $\Delta U_4 = 0$ and $|\Delta U_4| >> k_B T$ ($\Delta U_4 < 0$).

Comparing this expression to the difference in quadrupole-quadrupole interaction energies for the structures concerned, ΔH_{or}=-(9/8)$NU\sin^4\theta$, we find that with θ=25^0 the thermodynamic effect dominates over quadrupole at T>15 K. In the angle θ range of 24.6 to 37.5^0, the phase BD results, with the angle values $\varphi_1 = \varphi_2$ and $\varphi_3 = \varphi_4$ changing continuously (it is evident that the phase AC gives way to BD at a somewhat larger angle θ=30.6^0). At the point D (θ=37.5^0), azimuthal angles change jump-like to the same value, 45^0, and remain constant until the point F (θ=63.4^0) is reached; from this point on, they change continuously up to the values $\varphi_1 = \varphi_4 = 0$, $\varphi_2 = \varphi_3 = 90^0$ assumed at the point G (θ=90^0) which corresponds to the ground state of free quadrupoles (with $U(\theta_j, \varphi_j) = 0$). The energy of this absolute minimum is H_{or}=-6NU.

It is interesting to correlate the theoretically derived ground states with the experimentally observed orientational structure of CO molecules on the NaCl(100) surface. At temperatures below 25 K, this structure is characterized by the values θ=25^0 and $\varphi_1 = \varphi_2 = 0$, $\varphi_3 = \varphi_4 = 180^0$,[28,99] i.e. coincides with the above-considered phase AB. Its predominance over the phase AC may be attributed not only to the role of thermodynamic fluctuations but also to the effect of the adsorption potential which should include negative values of the parameter ΔU_4, if expressed as in Eq. (2.4.1). The value of this parameter estimated from the frequency of the low-frequency mode[99] $\omega_l \approx 37.5$ cm^{-1} (which causes the dephasing of the high-frequency vibration C-O) shows that the azimuthal component of the adsorption potential is characterized by energies of the order k_BT in the temperature region of the observable orientational phase AB and should therefore be taken into account.

Consider the approximation of four discrete molecular orientations along the axes of a square lattice ($|\Delta U_4| >> k_BT$, $\Delta U_4 < 0$). To conveniently describe orientations, we introduce, at each lattice site, two spin variables, $\sigma_{nm} = \pm 1$ and $s_{nm} = \pm 1$, which are related to unit vectors ξ_{nm} as follows (Fig. 2.18):

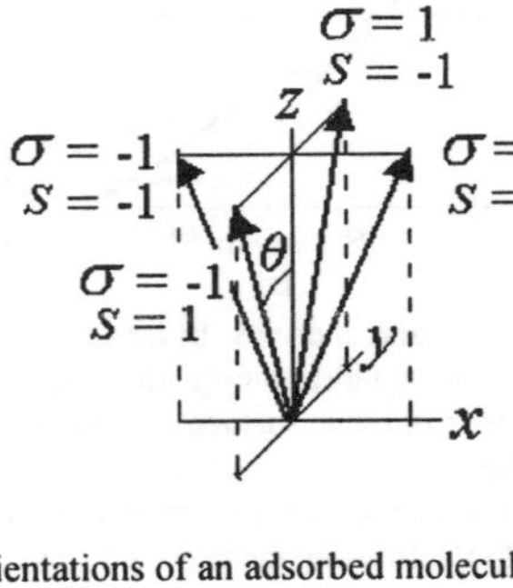

Fig. 2.18. Four discrete orientations of an adsorbed molecule inclined to a surface plane.

$$\xi_{nm} = \frac{1}{2}\left((-1)^n(\sigma_{nm}+s_{nm}),\ (-1)^m(\sigma_{nm}-s_{nm}),\ 0\right). \tag{2.4.4}$$

Substituting Eq. (2.4.4) into (2.4.2) leads to:

$$\begin{aligned} H_{\text{or}} = H_0 + \sum_{nm} \Big[& -J_1\left(\sigma_{nm}\sigma_{n,m+1} + s_{nm}s_{n,m+1} + \sigma_{nm}\sigma_{n+1,m} + s_{nm}s_{n+1,m}\right) \\ & - J_2\left(\sigma_{nm}s_{n,m+1} + s_{nm}\sigma_{n,m+1} - \sigma_{nm}s_{n+1,m} - s_{nm}\sigma_{n+1,m}\right) \\ & + J_3\left(\sigma_{nm}s_{nm}\sigma_{n,m+1}s_{n,m+1} + \sigma_{nm}s_{nm}\sigma_{n+1,m}s_{n+1,m}\right)\Big], \end{aligned} \tag{2.4.5}$$

where

$$\begin{aligned} H_0 &= N\left[2B_0 + 2B_1\sin^2\theta + \frac{1}{2}(B_2+2B_3+B_4+4B_5)\sin^4\theta \right. \\ &\qquad \left. + 2B_3\cos^4\theta + \frac{1}{2}\Delta U_4\right], \\ J_1 &= -\frac{1}{4}B_4\cos^2\theta\sin^2\theta \Rightarrow \frac{15}{4}U\cos^2\theta\sin^2\theta, \\ J_2 &= -\left(\frac{1}{4}B_4 + B_3\right)\cos^2\theta\sin^2\theta \Rightarrow \frac{9}{4}U\cos^2\theta\sin^2\theta, \\ J_3 &= \frac{1}{4}(B_2+2B_3+B_4)\sin^4\theta \Rightarrow \frac{57}{16}U\sin^4\theta. \end{aligned} \tag{2.4.6}$$

(Values of the coefficients J_1, J_2, J_3 for purely quadrupole-quadrupole interactions are marked by arrows). As far as $J_1 > J_2 > 0$ and $J_3 > 0$, ground states of the Hamiltonian expressed by Eq. (2.4.6) can be of two kinds:

$$\begin{aligned} H_{\text{or}} &= H_0 + N(-4J_1 + 2J_3), \qquad & \sigma_{nm} = s_{nm} = 1, \qquad & J_1 > J_3\,; \\ H_{\text{or}} &= H_0 - 2NJ_3, & \sigma_{nm}s_{nm} = (-1)^{n+m}, \qquad & J_1 < J_3\,. \end{aligned} \tag{2.4.7}$$

The phase AE in Fig. 2.17 which extends to the value $\theta = 45.7^0$ corresponds to a first-kind ground state. In the angle θ range of 45.7 to 90^0, a second-kind ground state is realized, at $\theta = 90^0$ it coinciding with the ground state of free quadrupoles on a square lattice.

Now we consider thermodynamic properties of the system described by the Hamiltonian (2.4.5); it is a generalized Hamiltonian of the isotropic Ashkin-Teller model[100,101] expressed in terms of interactions between pairs of spins σ_{nm} and s_{nm} which characterize each lattice site nm of a square lattice. Hamiltonian (2.4.5) differs from the known one in that it includes not only the contribution from the four-spin interaction (the term with the coefficient J_3), but also the anisotropic contribution (the term with the coefficient J_2) which accounts for cross interactions of spins σ_{nm} and s_{nm} between neighboring lattice sites. This term is so structured that it vanishes if there are no fluctuation interactions between σ- and s-subsystems. As a result, with sufficiently small coefficients J_2, we arrive at a typical phase diagram of the isotropic Ashkin-Teller model,[101,102] limited by the plausible values of coefficients in Eq. (2.4.6). At $J_1 > J_3$, the phase transition line

$$\exp(2K_3) = \sinh 2K_1, \quad K_i \equiv J_i/k_B T \tag{2.4.8}$$

(see the solid line on Fig. 2.19) separates the high-temperature region in which average spin variables and their products are equal to zero (disordered phase) from the low-temperature region where these variables assume nonzero positive values (ferromagnetically ordered phase).

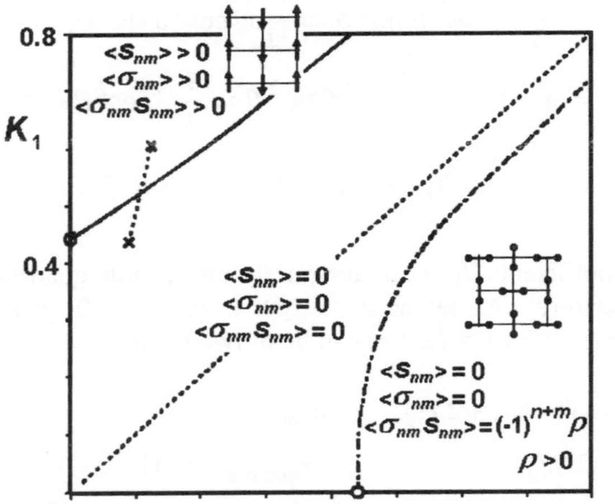

Fig. 2.19. A phase diagram of orientational states for adsorbed molecules on a square lattice. Phase-separating solid and dash-dotted lines correspond to the case $K_2 = 0$. The dotted line enclosed by the markers × specifies parameters of the system CO/NaCl(100) ($K_3/K_1 = 0.207$, $\theta = 25^0$)

This line starts at the point $K_1 \approx 0.4407$, $K_3 = 0$ corresponding to the critical point of the isotropic Ising model on a square lattice. At $K_3 \rightarrow K_1$, the ground-state structure changes and the critical temperature approaches zero. Further, at $J_1 < J_3$, the low-temperature phase is characterized by partial antiferromagnetic ordering, $<\sigma_{nm}s_{nm}>$ being non-zero (with alternating signs at neighboring lattice sites) but $<\sigma_{nm}>$ and $<s_{nm}>$ vanishing. The position of the critical line (the dash-dotted line in Fig. 2.19) separating the ordered phase from the disordered one is not precisely known but the point $K_1 = 0$, $K_3 \approx 0.4407$ again corresponds to the Ising isotropic model with the spin variables $\sigma_{nm}s_{nm}$. This point dictates the critical temperature for a system of molecules lying in the surface plane $(\theta = 90^0)$.

The dotted line in Fig. 2.19 enclosed by two markers × represents the ratio $K_3/K_1 = 0.207$ at the inclination angle $\theta = 25^0$ observed for the system CO/NaCl(100). The point at which this line intersects the solid one gives the critical temperature which proves equal to 20K if the value $U = 1.63$ meV is substituted. If dispersion and repulsion interactions estimated as in Table 2.3 are taken into account, the value T_c is shifted to 22 K, which agrees well with the experimental value $T_c = 17.5 \div 21.5$ K.[99]

2.5. Orientational phase transitions in planar systems of nonpolar molecules

A switch to double-angle vectors given by Eq. (2.3.12) not only significantly simplifies the treatment of orientation phase transitions in planar systems of nonpolar molecules but also leads to a number of substantial inferences on the transition nature. First of all note that the long-range-order parameter η (vanishing in a disordered phase and equal to unity at $T = 0$) in a d–dimensional space (specified by the orientations of long molecular axes) can be defined as:

$$\left\langle e_j^{\alpha} e_j^{\beta} \right\rangle = \frac{1}{d}(1-\eta)\delta_{\alpha\beta} + \xi_j^{\alpha}\xi_j^{\beta}\eta\,, \qquad (2.5.1)$$

where the unit vectors ξ_j determine the ground-state orientational structure. At $d = 3$ in the self-consistent-field approximation, the free energy of the system, if expanded in terms of the order parameter, can include a cubic term which gives rise to a first-order phase transition (e.g., like that described in α-form crystal structure for N_2).[47] Based on relations (2.5.1) and (2.3.12), and with allowance made for the properties of the matrices $\mathbf{S}_0$ and $\mathbf{S}_\gamma$ (see eqs. (2.3.13)-(2.3.15)), the long-range-order parameter can be written in the form typical of systems of polar molecules:

$$\langle \varepsilon_j \rangle = \eta \; \varepsilon_j \big|_{T=0} . \tag{2.5.2}$$

Denote the self-consistent field exerted on a molecule j by neighboring molecules as $\eta\widetilde{V}$ where $\widetilde{V}$ represents the absolute magnitudes of quantities (2.3.25) and (2.3.27) for various ground states of nonpolar molecules on square and triangular lattices. Then the equation for η appears as:

$$\eta = \frac{\sum\limits_{\alpha} \cos\alpha \exp(-\eta\widetilde{V}\cos\alpha / k_{\mathrm{B}}T)}{\sum\limits_{\alpha} \exp(-\eta\widetilde{V}\cos\alpha / k_{\mathrm{B}}T)} . \tag{2.5.3}$$

Here the summation is performed over all permissible orientations of the vector ε_j which is directed at the angle α to the vectors given by Eq. (2.5.2).

For an adsorption lattice of the C_{2n} symmetry, the angular part of the adsorption potential selects n preferable molecular orientations. It can be approximated by the hindered-rotation potential

$$U(\varphi_j) = \frac{1}{2}\Delta U_n \cos 2n\varphi_j \tag{2.5.4}$$

(for the symmetry C_3 and $n = 3$, the quantity $\cos 2n\varphi_j$ is replaced by $\cos 3\varphi_j$). Assume that the vector ε_j can have n symmetric orientations, so that $\alpha = 0, 2\pi/n, \ldots, 2\pi(n-1)/n$. The orientations of long molecular axes of nonpolar molecules will differ by the angle π/n but their total quantity will be again n, due to the equivalence of the orientations with the angles φ and $\varphi + \pi$. Expanding the right-hand side of Eq. (2.5.3) in terms of η, we arrive at the conclusion that a first-order phase transition[56,103,104] occurs only at $n = 3$, whereas a second-order phase transition results in all the other cases. Simplified self-consistency given by Eq. (2.5.3) allows only for the transitions between the ordered low-temperature phase (corresponding to the ground state) and the disordered high-temperature phase, without considering boundaries between various ordered phases (some of them may lie in an intermediate temperature range). For such boundaries on square and triangular lattices to be calculated, a more detailed treatment is needed.[85,86]

The phase transition temperature at $n \neq 3$ is specified by the relation:

$$k_B T_c = \begin{cases} \tilde{V}, & n = 2, \\ \tilde{V}/2, & n \geq 4. \end{cases} \tag{2.5.5}$$

It is noteworthy that the phase transition temperature falls by half on going from 2 to 4 possible orientations and then remains constant up to the $n \to \infty$ (the sums in expression (2.5.3) are replaced by the corresponding integrals). On the other hand, the self-consistent-field approximation is known to materially overestimate phase transition temperatures (especially since two-dimensional rather than three-dimensional systems are concerned). That is why, correlations with experimental data should involve more advanced statistical methods that take account of orientational fluctuation interactions. Unlike thoroughly studied two-dimensional systems with isotropic Heisenberg interactions and symmetry-breaking crystalline fields[105,106] which under certain conditions give rise to the short-range-ordered Berezinskii-Kosterlitz-Thouless phase,[107,108] systems with anisotropic potentials manifest a number of particular properties. To exemplify, strong anisotropy stabilizes long-range order in two-dimensional dipole systems even though the Goldstone mode may be exhibited in the orientational vibration spectrum.[46,66,67,70,109] At the same time, the two-dimensional orientational structures of quadrupoles under consideration do not display the Goldstone mode thus evidencing the long-range order in the low-temperature phase. The Monte Carlo simulation of orientational phase transitions for dipoles on a triangular lattice (with long-range dipole-dipole interactions) yields the following dependence of transition temperatures on the number of symmetric discrete orientations: $k_B T_c/(\mu^2/a^3) = 2.4$, 1.5, 1.08, and 0.92 at $n = 2$, 3, 4, and ∞, respectively.[110] The values T_c at $n = 4$ and ∞ differ from that at $n = 2$ by a factor of about two. An analogous simulation for a system of degenerate dipoles on a square lattice ($n = \infty$) provided the values $k_B T_c/(\mu^2/a^3) = 1.52 \pm 0.01$ for the short-range model and 0.75 with the long-range dipole forces included.[76,111] The first value is found to be close to the value 1.641 in an exactly solvable dipole short-range Ising model[75] in which the dipoles can have four discrete orientations along the diagonals of a square lattice.

Of particular interest are the models which enable statistical problems to be solved exactly. Let us suppose that the adsorption potential $U(\varphi_j)$ admits only two rigidly fixed molecular orientations ($n = 2$) with the angles $\varphi_j = 0$, $\pi/2$ for a square lattice and $\varphi_j = \pi/4$, $3\pi/4$ for a triangular lattice (Fig. 2.20 a and b); these requirements are respectively met at $|\Delta U_2| >> U$ with $\Delta U_2 < 0$ and at $\Delta U_2 > 0$. The orientations selected are compatible with the ground-state structure of quadrupoles on such lattices. Introducing "spin variables" $\sigma_{mn} = \pm 1$ (as in section 2.3, integer mn pairs label lattice sites) and expressing double-angle vectors as

$\boldsymbol{\varepsilon}_{mn} = (\sigma_{mn}, 0, 0)$ for a square lattice and $\boldsymbol{\varepsilon}_{mn} = (0, \sigma_{mn}, 0)$ for a triangular lattice, the Hamiltonian (2.3.17) can be represented for these lattices in the following form:

$$H_{or} = \left(\pm \frac{1}{4} \Delta U_2 + V_0 \right) N + \sum_{mn} \begin{cases} (D_1 + D_2)(\sigma_{mn}\sigma_{m,n+1} + \sigma_{mn}\sigma_{m+1,n}), \\ \left[D_1 \sigma_{mn}\sigma_{m,n+1} + \left(D_1 + \frac{3}{4} D_2 \right)(\sigma_{mn}\sigma_{m+1,n} + \sigma_{m,n+1}\sigma_{m+1,n}) \right], \end{cases} \tag{2.5.6}$$

where V_0 is given by formula (2.3.18) with $\left(\hat{\varepsilon}_j^x\right)^2 = 1$.

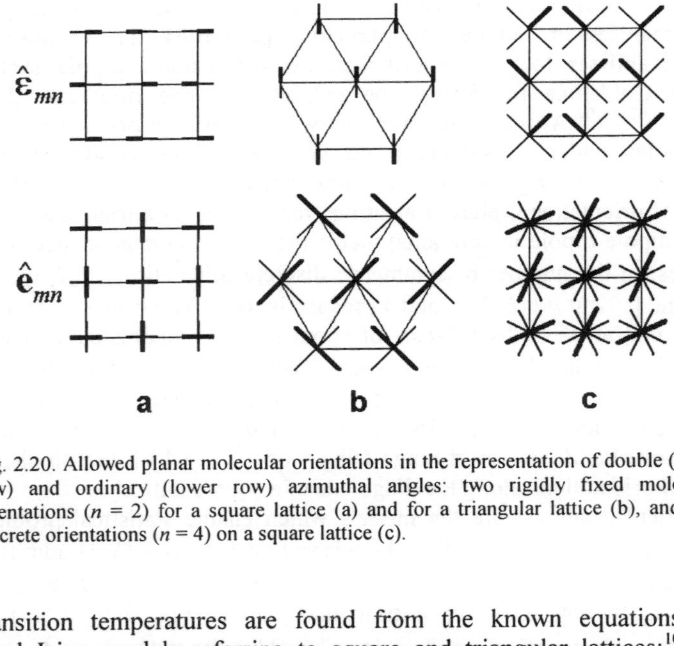

Fig. 2.20. Allowed planar molecular orientations in the representation of double (upper row) and ordinary (lower row) azimuthal angles: two rigidly fixed molecular orientations ($n = 2$) for a square lattice (a) and for a triangular lattice (b), and four discrete orientations ($n = 4$) on a square lattice (c).

Phase transition temperatures are found from the known equations for two-dimensional Ising models referring to square and triangular lattices;[101] with the parameters of the quadrupole potential, their solutions appear as:

$$k_B T = \begin{cases} 8.084U - \text{square lattice}, \\ 8.2474U - \text{triangular lattice}. \end{cases} \tag{2.5.7}$$

(Compare these values with 14.25 U and 13.6875 U respectively provided for the two lattice types by the self-consistent-field approximation (2.5.5) at $n = 2$).

It might be expected that at $n \geq 4$ the solutions of the Ising-model equations are half as large as values (2.5.7) in consequence of the changed dimensionality of the orientation space (see Eqs. (2.5.3) and (2.5.5)). To verify this conjecture, we turn to the case of four discrete orientations $\varphi_j = \pi/8, 3\pi/8, 5\pi/8, 7\pi/8$ ($n = 4$ and $\Delta U_2 >> U$ in Eq. (2.5.4)) on a square lattice (Fig. 2.20 c). The angle values are selected so that possible orientations of the double-angle vector $\hat{\boldsymbol{\varepsilon}}_j$ are parallel to the diagonals of the lattice site square and one can take advantage of the previously obtained[75] exact solution for the short-range dipole Hamiltonian. At the same time, the discrete orientations chosen do not coincide with the ground-state molecular orientations, which enables monitoring an interesting phenomenon, *viz.* stepwise reorganization of the orientational structure occurring with the lowering temperature.

To treat the case $n = 4$, introduce a couple of "spin variables" $\sigma_{mn} = \pm 1$ and $s_{mn} = \pm 1$ with the double-angle vectors $\hat{\boldsymbol{\varepsilon}}_{mn} = \left(1/\sqrt{2}\right)\left(\sigma_{mn}, s_{mn}, 0\right)$. Then the Hamiltonian (2.3.17) appears as:

$$H_{or} = \left(-\frac{1}{2}\Delta U_4 + V_0\right)N$$
$$+\frac{1}{2}\sum_{mn}\left[\left(D_1 + D_2\right)\left(\sigma_{mn}\sigma_{m,n+1} + \sigma_{mn}\sigma_{m+1,n}\right) + D_1\left(s_{mn}s_{m,n+1} + s_{mn}s_{m+1,n}\right)\right]. \qquad (2.5.8)$$

where V_0 is given by formula (2.3.18) with $\left(\hat{\varepsilon}_j^x\right)^2 = 1/2$. The ground state for quadrupole interactions implies that $\sigma_{mn} = (-1)^{m+n}$ and $s_{mn} = 1$, and has a lower energy than were the case without certain orientations selected. The average length of the vector $\hat{\boldsymbol{\varepsilon}}_j$ is characterized by two order parameters, η_x и η_y, respectively referring to σ- and s-subsystems:

$$\left\langle \hat{\boldsymbol{\varepsilon}}_{mn} \right\rangle = \frac{1}{\sqrt{2}}\left((-1)^{m+n}\eta_x, \eta_y, 0\right). \qquad (2.5.9)$$

With rising temperature, the values η_x and η_y, and hence $\left|\left\langle \hat{\boldsymbol{\varepsilon}}_{mn} \right\rangle\right|$ fall off to the first phase transition temperature:

$$k_B T_y = 3.404U \tag{2.5.10}$$

at which the order parameter η_y vanishes (see Fig. 2.21). As temperature increases further, the vector $\langle \hat{\varepsilon}_{mn} \rangle$ becomes oriented along the horizontal axis of the square lattice, whereas the average positions of molecular axes at the neighboring lattice sites form T-like structures just as in the ground state of a quadrupole system in the absence of selected orientations. If temperature goes still higher, the values η_x drop from 0.88 to zero, the vanishing order parameter being associated with the second phase transition occurring at the point

$$k_B T_x = 4.042U\,. \tag{2.5.11}$$

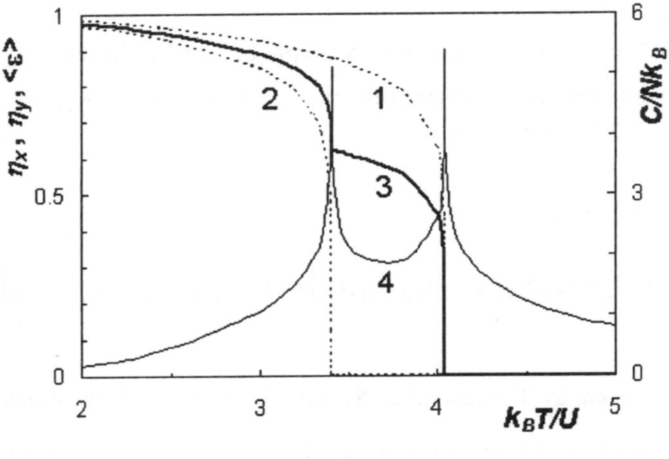

Fig. 2.21. Temperature dependences of the order parameters η_x (1), η_y (2), $\langle \hat{\varepsilon}_{mn} \rangle$ (3) and the specific heat C/Nk_B (4, the right-hand ordinate axis) for the model with four discrete orientations.

It is noteworthy that the critical value given by Eq. (2.5.11) is exactly half as large as for two quadrupole orientations on a square lattice (cf. Eq. (2.5.7)). This is a vindication of the inference that a halved critical temperature results from a corresponding change in the orientation space dimensionality. Thus, one might with good reason anticipate that the critical temperature for a triangular lattice of quadrupoles with arbitrary planar orientations should also be approximately half as large as that in relation (2.5.7), i.e., $k_B T_c \approx 4.1U$. Referring to the Monte Carlo calculation for 36 quadrupoles constrained to a plane in a triangular lattice[90], the critical temperature is estimated as $k_B T_c \approx 2.8U$. Improved Monte Carlo simulations for a triangular lattice with 400, 1600, 6400, and 10000 sites[83] took into account only the second term in Eq. (2.3.21) and thus underestimated the transition temperature so that $k_B T_c \approx 2.5U$. It should be pointed out that the value T_c for $n = \infty$ is now lowered somewhat more in reference to that for $n = 4$ than it is in the case of a ferroelectric structure of dipole moments on a square lattice.[109,110]

An orientation phase transition in a molecular monolayer N_2 on graphite surface was registered in the course of heat-capacity study[112] and neutron-diffraction study[113] at about $T_c = 30$ K. Low-energy electron diffraction experiments[5] show that the superlattice intensity decreases significantly near 30 K but persists up to 40 K. For a triangular lattice formed by molecular centers on graphite surface, the lattice constant value is well defined: $a \approx 4.26$ Å, whereas there is some disagreement on quadrupole moment values:[114] $Q=(1.42\pm0.08)\times10^{-26}$ esu cm^2 for an isolated nitrogen molecule,[115] 1.34×10^{-26} esu cm^2 in the framework of the Etters model,[116] and 1.173×10^{-26} esu cm^2 according to the X1 model.[117] The parameter U for these values can vary in the range from 0.61 to 0.87 meV; the value T_c accordingly varies in the range from 29 to 42 K at $n = 4$ ($k_B T_c \approx 4.1U$) and from 20 to 29 K at $n = \infty$ ($k_B T_c \approx 2.8U$). Provided a finite rotation barrier of the C_4 symmetry, transitions temperatures should assume intermediate values between the two intervals indicated, which is the case in experimental studies.

Interestingly, the orientation phase transition in question has been treated in a record great number of publications (see surveys [2-4] and references cited therein). Especially its order received much space in the literature. Using the histogram method of Monte Carlo simulation combined with an analysis of finite-size scaling and the fourth-order energy cumulant, Cai[118] earnestly showed a continuous second-order transition in this system, contrary to the previous simulation study[83] and renormalization-group analysis[119] that suggested a first-order transition. The critical exponents[118] are determined to be those of the three-state Potts model.[120] One can, however, judge from the discussion in Ref. 3 that the order of this phase transition is still an open question.

Chapter 3

Dispersion laws for collective excitations on complex planar lattices

In an effort to understand the mechanisms involved in formation of complex orientational structures of adsorbed molecules and to describe orientational, vibrational, and electronic excitations in systems of this kind, a new approach to solid surface theory has been developed which treats the properties of two-dimensional dipole systems.[61,109,121] In adsorbed layers, dipole forces are the main contributors to lateral interactions both of dynamic dipole moments of vibrational or electronic molecular excitations and of static dipole moments (for polar molecules). In the previous chapter, we demonstrated that all the information on lateral interactions within a system is carried by the Fourier components of the dipole-dipole interaction tensors. In this chapter, we consider basic spectral parameters for two-dimensional lattice systems in which the unit cells contain several inequivalent molecules. As seen from Sec. 2.1, such structures are intrinsic in many systems of adsorbed molecules. For the Fourier components in question, the lattice-sublattice relations will be derived which enable, in particular, various parameters of orientational structures on a complex lattice to be expressed in terms of known characteristics of its Bravais sublattices. In the framework of such a treatment, the ground state of the system concerned as well as the infrared-active spectral frequencies of valence dipole vibrations will be elucidated.

3.1. Complex orientational structures and corresponding Davydov-split spectral lines

Let us consider an arbitrary planar lattice that consists of n Bravais sublattices having unit cell $\mathbf{A}_1$, $\mathbf{A}_2$. Then the sites of the complex lattice are given by the vectors $\mathbf{R} + \mathbf{r}_j$, where $\mathbf{R} = m_1\mathbf{A}_1 + m_2\mathbf{A}_2$ specifies the sites of the basic Bravais sublattice and $\mathbf{r}_j$ characterizes the positions of the jth sublattice sites within the unit cell defined above (Fig. 3.1). Let each site of the complex lattice be occupied by an adsorbed molecule having dipole moment $\mu_{\mathbf{R}j} = (\mu_j + q_j x_{\mathbf{R}j})\mathbf{e}_{\mathbf{R}j}$. For the sake of simplicity, here we consider stretching (valence) vibrations of the dynamic dipole moment $q_j x_{\mathbf{R}j}\mathbf{e}_{\mathbf{R}j}$ (q_j is an effective charge; $x_{\mathbf{R}j}$ is a radial shift) along the static dipole moment $\mu_j\mathbf{e}_{\mathbf{R}j}$ ($\mathbf{e}_{\mathbf{R}j}$ is a unit vector).[122] At the end of the section, the approach developed will also be extended to bending vibrations. The total Hamiltonian of the system under consideration can be represented as a sum:

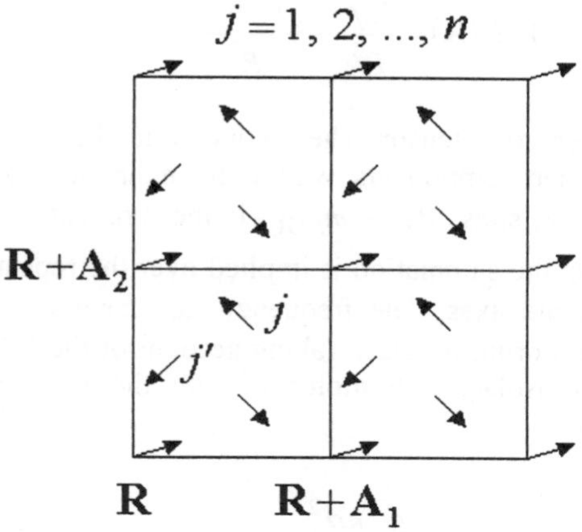

Fig. 3.1. A planar lattice consisting of *n* Bravais sublattices of identical oriented molecules.

$$H = H_0 + H_{\text{ex}} \tag{3.1.1}$$

contributed by static dipole-dipole interactions

$$\begin{aligned} H_0 &= \frac{1}{2} \sum_{\mathbf{R}j,\mathbf{R}'j'} \mu_j \mu_{j'} V^{\alpha\beta}\left(\mathbf{R} - \mathbf{R}' + \mathbf{r}_{jj'}\right) e^{\alpha}_{\mathbf{R}j} e^{\beta}_{\mathbf{R}'j'} \\ &= \frac{1}{2} N_0 \sum_{\mathbf{K}jj'} \mu_j \mu_{j'} \tilde{V}^{\alpha\beta}_{jj'}(\mathbf{K}) e^{\alpha}_j(-\mathbf{K}) e^{\beta}_{j'}(\mathbf{K}) \end{aligned} \tag{3.1.2}$$

and the Hamiltonian of vibrational excitations

$$\begin{aligned} H_{\text{ex}} &= \sum_{\mathbf{R}j} \left(\frac{P^2_{\mathbf{R}j}}{2m_j} + \frac{1}{2} m_j \omega^2_{\mathbf{R}j} x^2_{\mathbf{R}j} \right) \\ &+ \frac{1}{2} \sum_{\mathbf{R}j,\mathbf{R}'j'} q_j q_{j'} V^{\alpha\beta}\left(\mathbf{R} - \mathbf{R}' + \mathbf{r}_{jj'}\right) e^{\alpha}_{\mathbf{R}j} e^{\beta}_{\mathbf{R}'j'} x_{\mathbf{R}j} x_{\mathbf{R}'j'} . \end{aligned} \tag{3.1.3}$$

Here

$$V^{\alpha\beta}(\mathbf{R}) = \frac{\delta_{\alpha\beta}}{R^3} - 3\frac{R_\alpha R_\beta}{R^5} \tag{3.1.4}$$

is a dipole-dipole interaction tensor. The second addend in Eq. (3.1.2) corresponds to switching to Fourier components within the main area of the basic Bravais sublattice containing N_0 sites; $P_{\mathbf{R}j} = m_j \dot{x}_{\mathbf{R}j}$ is the momentum of the vibration $x_{\mathbf{R}j}$ with reduced mass m_j; the summation is implied over the repeated Greek indices of the Cartesian coordinate axes. The frequency $\omega_{\mathbf{R}j}$ depends on the static electric fields caused by neighboring dipoles. Taking account of the cubic anharmonicity α_j for vibrations of the isolated jth molecule with the force constant k_j and the frequency specified as

$$\omega_{0j} = \left(\frac{k_j}{m_j}\right)^{1/2} - \frac{5}{6}\frac{\hbar\alpha_j^2}{m_j k_j^2}, \tag{3.1.5}$$

we are led to the static frequency renormalization:[61,121]

$$\omega_{\mathbf{R}j}^2 = \omega_{0j}^2 + \frac{2\alpha_j q_j}{m_j k_j}\sum_{\mathbf{R}'j'} \mu_{j'} V^{\alpha\beta}\left(\mathbf{R} - \mathbf{R}' + \mathbf{r}_{jj'}\right) e_{\mathbf{R}j}^{\alpha} e_{\mathbf{R}'j'}^{\beta}. \tag{3.1.6}$$

Now we consider the two principal challenges presented by the systems under study. The first involves determination of the ground state for a lattice system of static dipole moments and implies H_0 minimization over all possible orientations of the vectors $\mathbf{e}_{\mathbf{R}j}$. Molecules are assumed to have uniform dipole moments, $\mu_j = \mu$, with arbitrary orientations, $\mathbf{e}_{\mathbf{R}j}$, in the absence of dipole-dipole interactions. On finding eigenvalues and eigenvectors of the tensor $\widetilde{V}_{jj'}^{\alpha\beta}(\mathbf{K})$ by the equation

$$\sum_{j'} \widetilde{V}_{jj'}^{\alpha\beta}(\mathbf{K}) C_{j'p}^{\beta\nu}(\mathbf{K}) = V_p^{\nu}(\mathbf{K}) C_{jp}^{\alpha\nu}(\mathbf{K}), \quad j,\ p = 1,\ \ldots,\ n; \quad \nu = 1,\ 2,\ 3, \tag{3.1.7}$$

the quadratic form (3.1.2) can be presented as a sum of squares

$$H_0 = \frac{1}{2} N\mu^2 \sum_{\mathbf{K},p,\nu} V_p^{\nu}(\mathbf{K}) \left|\eta_p^{\nu}(\mathbf{K})\right|^2, \quad N = nN_0, \quad \sum_{\mathbf{K},p,\nu} \left|\eta_p^{\nu}(\mathbf{K})\right|^2 = 1, \tag{3.1.8}$$

$$e_{\mathbf{R}j}^{\alpha} = \sqrt{n} \sum_{\mathbf{K},p,\nu} C_{jp}^{\alpha\nu}(\mathbf{K}) \eta_p^{\nu}(\mathbf{K}) \exp\left[i\mathbf{K}\cdot\left(\mathbf{R}+\mathbf{r}_j\right)\right]. \tag{3.1.9}$$

As is now evident, the ground-state energy is determined by the deepest minimum among those inherent in a family of functions $V_p^{\nu}(\mathbf{K})$, and the corresponding dipole moment configuration $\mathbf{e}_{\mathbf{R}j}$ is specified by the eigenvectors $C_{jp}^{\alpha\nu}(\mathbf{K})$ (see Eq. (3.1.9)). The above procedure for finding the ground state of a dipole system on a complex lattice is a generalization of the technique used previously for simple Bravais lattices (see Sec. 2.2).

The second problem of interest is to find normal vibrational frequencies and integral intensities for spectral lines that are active in infrared absorption spectra. In this instance, we can consider the molecular orientations, $\mathbf{e}_{\mathbf{R}j}$, to be already specified. Further, it is of no significance whether the orientational structure $\mathbf{e}_{\mathbf{R}j}$ results from energy minimization for static dipole-dipole interactions or from the competition of any other interactions (e.g. adsorption potentials). For non-polar molecules ($\mu_j = 0$), the vectors $\mathbf{e}_{\mathbf{R}j}$ describe dipole moment orientations for dipole transitions.

Our concern here is with the periodic orientations of dynamic dipole moments, when a unit cell of a two-dimensional crystal may be thought of as containing n orientationally inequivalent molecules:

$$e_{\mathbf{R}j}^{\alpha} = e_j^{\alpha}, \quad \omega_{\mathbf{R}j}^2 = \omega_j^2, \quad j = 1, \ldots, n. \tag{3.1.10}$$

In passing from the first to the second problem, a feature of importance should be borne in mind. The periods of the orientational structure (3.1.9) can exceed those of the basic Bravais sublattice, $\mathbf{A}_1$, $\mathbf{A}_2$. If this is the case, the unit cell $\mathbf{A}_1$, $\mathbf{A}_2$ should be enlarged so that conditions (3.1.10) can be met and translations onto the new vectors $\mathbf{R}$ can reproduce the orientations of adsorbed molecules. Then the excitation Hamiltonian (3.1.3) can be represented in the Fourier form with respect to the wave-vector $\mathbf{K}$ as

$$H_{\text{ex}} = \sum_{\mathbf{K}} H_{\text{ex}}(\mathbf{K}),$$

$$H_{\text{ex}}(\mathbf{K}) = \sum_j \left(\frac{\left|\widetilde{P}_{\mathbf{K}j}\right|^2}{2m_j} + \frac{1}{2} m_j \omega_j^2 \left|\widetilde{x}_{\mathbf{K}j}\right|^2 \right) + \frac{1}{2} \sum_{j,j'} q_j q_{j'} \widetilde{V}_{jj'}^{\alpha\beta}(\mathbf{K}) e_j^{\alpha} e_{j'}^{\beta} \widetilde{x}_{\mathbf{K}j} \widetilde{x}_{-\mathbf{K}j'} \tag{3.1.11}$$

and it is possible to introduce the normal coordinates ξ_l for the mode of interest (**K** = 0) arising in infrared spectra (for a detailed transform to normal coordinates in a system of bound harmonic oscillators see Appendix 1):

$$\widetilde{x}_{0j} = m_j^{-1/2}\sum_l S_{jl}\xi_l, \quad \widetilde{P}_{0j} = m_j^{1/2}\sum_l S_{jl}\dot{\xi}_l . \tag{3.1.12}$$

Here the unitary matrices S_{jl}

$$\sum_j S_{jl}S^*_{jl'} = \delta_{ll'}, \quad \sum_l S_{jl}S^*_{j'l} = \delta_{jj'} \tag{3.1.13}$$

are constructed from the eigenvectors of the squared frequency matrix:

$$\Phi_{jj'} = \omega_j^2\delta_{jj'} + \frac{q_j q_{j'}}{\sqrt{m_j m_{j'}}}\widetilde{V}^{\alpha\beta}_{jj'}(0)e^\alpha_j e^\beta_{j'} , \tag{3.1.14}$$

$$\omega_j^2 = \omega_{0j}^2 + \frac{2\alpha_j q_j}{m_j k_j}e^\alpha_j \sum_{j'}\widetilde{V}^{\alpha\beta}_{jj'}(0)e^\beta_{j'} . \tag{3.1.15}$$

The eigenvalues of this matrix satisfying the equation

$$\sum_{j'}\Phi_{jj'}S_{j'l} = \Omega_l^2 S_{jl} \tag{3.1.16}$$

determine the squared frequencies of normal vibrations:

$$H_{\text{ex}}(0) = \frac{1}{2}\sum_l\left(\left|\dot{\xi}_l\right|^2 + \Omega_l^2\left|\xi_l\right|^2\right). \tag{3.1.17}$$

The coefficient of infrared light absorption by an adsorbed molecular monolayer of the surface concentration N/F depends both on the angle of incidence ϑ reckoned from a normal to the surface and on the effective absorption cross-section $\sigma(\omega)$:[122]

$$A(\omega) = \frac{N}{F\cos\vartheta}\sigma(\omega), \tag{3.1.18}$$

where $\sigma(\omega)$ is equal to the ratio of the power absorbed by a molecule, $Q(\omega)$, to the energy flow density for an incident wave averaged over its vibrational period:

$$\sigma(\omega)=\frac{8\pi}{c_0 E^2}Q(\omega), \quad Q(\omega)=\frac{1}{2}\omega\widetilde{E}^{\alpha}\widetilde{E}^{\beta}\frac{1}{n}\sum_{jj'}\operatorname{Im}\widetilde{\chi}_{jj'}^{\alpha\beta}(0,\omega). \tag{3.1.19}$$

Here $\widetilde{E}$ and E are electric field amplitudes on the surface and in vacuo interrelated by the Fresnel formulae; c_0 is the velocity of light in vacuo; $\widetilde{\chi}_{jj'}^{\alpha\beta}(\mathbf{K},\omega)$ is a susceptibility tensor defined by the equation:

$$\widetilde{\chi}_{jj'}^{\alpha\beta}(\mathbf{K},\omega)=\widetilde{\chi}_{j}^{\alpha\beta}(\omega)\delta_{jj'}-\widetilde{\chi}_{j}^{\alpha\gamma}(\omega)\sum_{j''}V_{jj''}^{\gamma\lambda}(\mathbf{K})\widetilde{\chi}_{j''j'}^{\lambda\beta}(\mathbf{K},\omega) \tag{3.1.20}$$

with the following form of the polarizability tensor for the jth molecule:

$$\widetilde{\chi}_{j}^{\alpha\beta}(\omega)=-\frac{q_j^2}{m_j}\frac{e_j^{\alpha}e_j^{\beta}}{\omega^2-\omega_j^2+i0\operatorname{sign}\omega}. \tag{3.1.21}$$

The solution of Eq. (3.1.20) at $\mathbf{K}=0$ is expressed in terms of eigenvalues and eigenvectors specified by Eq. (3.1.16):

$$\widetilde{\chi}_{jj'}^{\alpha\beta}(\omega)=-\frac{q_j q_{j'}}{\sqrt{m_j m_{j'}}}e_j^{\alpha}e_{j'}^{\beta}\sum_{l}\frac{S_{jl}S_{j'l}^{*}}{\omega^2-\Omega_l^2+i0\operatorname{sign}\omega}. \tag{3.1.22}$$

Substituting Eq. (3.1.22) into formulae (3.1.18) and (3.1.19) leads to the expression for the integral intensity of the lth spectral line:

$$A_l=\int_{\Omega_l-0}^{\Omega_l+0}A(\omega)d\omega=\frac{2\pi^2 N}{nc_0 F\cos\vartheta}\left|\sum_j q_j m_j^{-1/2}(\boldsymbol{\varepsilon}\cdot\mathbf{e}_j)S_{jl}\right|^2, \tag{3.1.23}$$

where $\boldsymbol{\varepsilon}=\widetilde{\mathbf{E}}/E$ specifies a vector of the electric field orientation on the surface in units of electric field amplitude for infrared radiation in vacuo. Thus, n molecules in the unit cell of a two-dimensional crystal are characterized by n spectral lines. If all the molecules are identical (i.e., q_j, m_j, and ω_{0j} are independent of j) and exposed to the same static electric fields generated by their dipole environment (i.e., ω_j is independent of j), whereas their orientations $\mathbf{e}_j$ are different, then formation of a crystal is accompanied by the splitting of a non-degenerate vibrational state for each free molecule into several differently polarized normal vibrations, which phenomenon is known as Davydov splitting.

The above relations can easily be generalized to the case when each molecule has several vibrational modes, e.g., displays not only stretching but bending vibrations as well.[123] To classify these intramolecular modes, introduce the additional index v and the vibrational polarizability $\chi_{jv} = q_{jv}^2 / m_{jv}\omega_{jv}^2$ for the mode v of the jth molecule. Thus, the unit vectors $\mathbf{e}_{jv}$ for the orientations of the corresponding vibrational displacements are now characterized by the couple of the indices j and v. Expression (3.1.14) for the squared frequency matrix accordingly becomes:

$$\Phi_{jv,j'v'} = \omega_{jv}^2 \delta_{jj'}\delta_{vv'} + \omega_{jv}\omega_{j'v'}\sqrt{\chi_{jv}\chi_{j'v'}}\,\tilde{V}_{jj'}^{\alpha\beta}(0)e_{jv}^{\alpha}e_{j'v'}^{\beta}. \qquad (3.1.24)$$

As a result, the indices jv and $j'v'$ should be replaced for j and j' in Eq. (3.1.16), and the expression for the integral intensity of the lth spectral line should include additional summation over v:

$$A_l = \int_{\Omega_l - 0}^{\Omega_l + 0} A(\omega)d\omega = \frac{2\pi^2 N}{nc_0 F\cos\vartheta}\left|\sum_{jv}\omega_{jv}\sqrt{\chi_{jv}}\left(\boldsymbol{\varepsilon}\cdot\mathbf{e}_{jv}\right)S_{jv,l}\right|^2. \qquad (3.1.25)$$

Polarization Fourier transform infrared surface spectroscopy makes use of light beams polarized in two mutually perpendicular directions (Fig. 3.2), surface-parallel (s-polarization) and surface-normal (p-polarization).

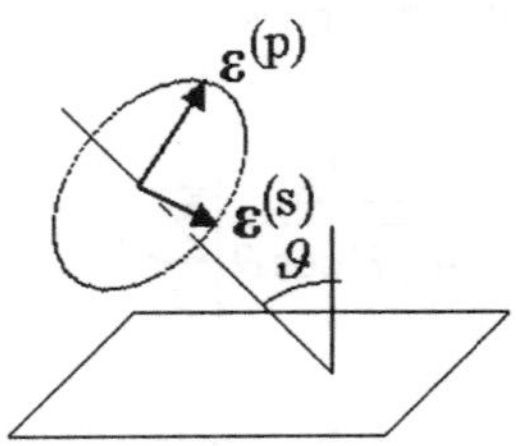

Fig. 3.2. Orientations of p- and s-polarized light beam with the angle of incidence ϑ.

Since adsorbed molecules are exposed both to the incident and reflected light waves whose electric field amplitudes are interrelated by the Fresnel formulae, the vectors in s- and p-polarizations appear as:

$$\varepsilon^{(s)} = 1 + r^{(s)}, \quad \varepsilon_{\parallel}^{(p)} = \left(1 - r^{(p)}\right)\cos\vartheta, \quad \varepsilon_{\perp}^{(p)} = \left(1 + r^{(p)}\right)\sin\vartheta,$$
$$r^{(s)} = -\frac{\sin(\vartheta - \vartheta_S)}{\sin(\vartheta + \vartheta_S)}, \quad r^{(p)} = \frac{\tan(\vartheta - \vartheta_S)}{\tan(\vartheta + \vartheta_S)}, \quad \sin\vartheta = n_S \sin\vartheta_S, \tag{3.1.26}$$

where n_S is the refractive index of the surface material (with the refractive angle ϑ_S); $r^{(s)}$ and $r^{(p)}$ denote the reflection coefficients for s- and p-polarized radiation. Measuring integral intensities in the two polarization furnishes significant evidence on the adsorbate orientational structure.

3.2. Lattice-sublattice relations and their application to orientational structures on square and honeycomb lattices

In both problems considered in Section 3.1, orientational structures and basic spectroscopic parameters are governed by the Fourier components of the dipole-dipole intersublattice interaction tensor $\tilde{V}_{jj'}^{\alpha\beta}(\mathbf{K})$. The behavior of the quantities $\tilde{V}_{jj}^{\alpha\beta}(\mathbf{K})$ for an arbitrary Bravais sublattice has been studied rather thoroughly.[46,61,121] Thus, the challenge is to calculate most efficiently the values of $\tilde{V}_{jj'}^{\alpha\beta}(\mathbf{K})$ at $j' \neq j$. We show that such intersublattice interactions are related to the known tensors $\tilde{V}_{jj}^{\alpha\beta}(\mathbf{K})$ for the basic Bravais sublattice. To do this, let us construct a denser Bravais lattice whose site set $\mathbf{r} = m_1\mathbf{a}_1 + m_2\mathbf{a}_2$ contains that of the complex lattice under study. Naturally, construction of this kind is not always possible; however, in the most interesting case presented by symmetric lattices, the basis vectors for the basic Bravais lattice are expressible as integer-coefficient linear combinations of vectors for the dense lattice: $\mathbf{A}_j = n_{j1}\mathbf{a}_1 + n_{j2}\mathbf{a}_2$ (j = 1, 2; $n_{jj'}$ are integers). Then the unit cell areas for these lattices are in the $\det\hat{n} = n$ relationship to each other, and the basis vectors for reciprocal lattices satisfying the equations $\mathbf{A}_j \cdot \mathbf{B}_{j'} = 2\pi\delta_{jj'}$ and $\mathbf{a}_j \cdot \mathbf{b}_{j'} = 2\pi\delta_{jj'}$ are mutually expressible as rational-coefficient linear combinations:

$$\mathbf{B}_j = \sum_{j'} \left(\hat{n}^{-1}\right)_{j'j} \mathbf{b}_{j'} \,. \tag{3.2.1}$$

By introducing the Fourier components $\widetilde{V}^{\alpha\beta}(\mathbf{K})$ of the dipole-dipole interaction tensor for the dense Bravais lattice, the desired quantities $\widetilde{V}^{\alpha\beta}_{jj'}(\mathbf{K})$ for the complex lattice can be determined as[122]

$$\widetilde{V}^{\alpha\beta}_{jj'}(\mathbf{K}) = \frac{1}{n} \sum_{\mathbf{B}} \widetilde{V}^{\alpha\beta}(\mathbf{K}+\mathbf{B}) \exp\left(i\mathbf{B}\cdot\mathbf{r}_{jj'}\right), \tag{3.2.2}$$

where summation is performed over all the integer-coefficient linear combinations of the vectors $\mathbf{B}_1$ and $\mathbf{B}_2$ falling within the first Brillouin zone of the dense Bravais lattice. The relation (3.2.2) proves very convenient both in elucidating the relationships between tensors $\widetilde{V}^{\alpha\beta}(\mathbf{K})$ at different points of the first Brillouin zone and in calculating intersublattice interactions within the complex lattice. Lattice-sublattice relations (3.2.2) can also be generalized to arbitrary-rank interaction tensors descending with distance by the power law with an arbitrary power.[109]

In the case of a square lattice, only two values of the lattice sums are necessary:

$$D_{\mathrm{F}} = -4.516811, \quad D_{\mathbf{A}} = -5.098873\,. \tag{3.2.3}$$

These determine the energies $H_0 = (1/2)N_0(\mu^2/a^3)D_{\mathrm{F,A}}$ (a is a lattice constant) for a dipole system with ferroelectric or antiferroelectric ordering in the lattice plane (see Fig. 3.3).[56]

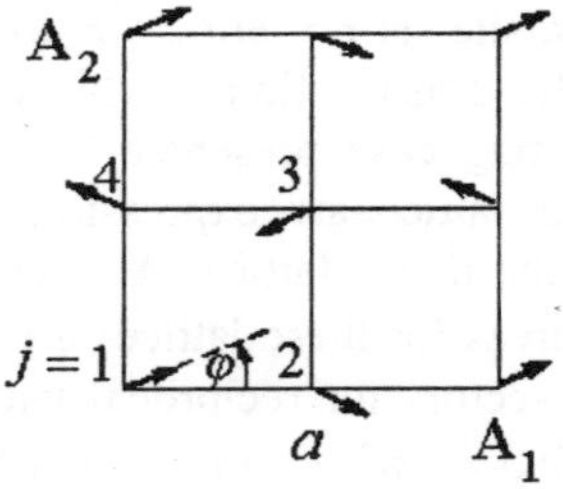

Fig. 3.3. The ground-state configuration of dipoles on a square lattice and the numbering of sublattices with the same dipole orientations.

Thus, the nonzero values of $D^{\alpha\beta}(\mathbf{K})=a^3\widetilde{V}^{\alpha\beta}(\mathbf{K})$ at the symmetric points of the first Brillouin zone and $D^{\alpha\beta}_{jj'}(\mathbf{K})=a^3\widetilde{V}^{\alpha\beta}_{jj'}(\mathbf{K})$, with j labelling sublattices as indicated in Fig. 3.3, are equal to:

$$
\begin{gathered}
\hat{D}(0)=D_{\mathrm{F}}\hat{\mathbf{\Gamma}},\ D^{xx}\left(\frac{1}{2}\mathbf{b}_1\right)=\frac{1}{2}\left(1-\sqrt{2}\right)D_{\mathrm{F}}-D_{\mathrm{A}},\ D^{yy}\left(\frac{1}{2}\mathbf{b}_1\right)=D_{\mathrm{A}},\\
D^{zz}\left(\frac{1}{2}\mathbf{b}_1\right)=-\frac{1}{2}\left(1-\sqrt{2}\right)D_{\mathrm{F}},\ \hat{D}\left[\frac{1}{2}(\mathbf{b}_1+\mathbf{b}_2)\right]=\left(2^{-1/2}-1\right)D_{\mathrm{F}}\hat{\mathbf{\Gamma}},\\
\hat{D}_{jj}(0)=\frac{1}{8}D_{\mathrm{F}}\hat{\mathbf{\Gamma}},\ \hat{D}_{13}(0)=\hat{D}_{24}(0)=\frac{1}{8}\left(2\sqrt{2}-1\right)D_{\mathrm{F}}\hat{\mathbf{\Gamma}},\\
\hat{D}_{14}(0)=\hat{D}_{23}(0)=\frac{1}{4}\left\{\left(2-2^{-1/2}\right)D_{\mathrm{F}}\hat{\mathbf{\Gamma}}+\left[\frac{1}{2}\left(1-\sqrt{2}\right)D_{\mathrm{F}}-2D_{\mathrm{A}}\right]\hat{\mathbf{\Lambda}}\right\},\\
\hat{D}_{12}(0)=\hat{D}_{34}(0)=\frac{1}{4}\left\{\left(2-2^{-1/2}\right)D_{\mathrm{F}}\hat{\mathbf{\Gamma}}-\left[\frac{1}{2}\left(1-\sqrt{2}\right)D_{\mathrm{F}}-2D_{\mathrm{A}}\right]\hat{\mathbf{\Lambda}}\right\},\\
\hat{\mathbf{\Gamma}}=\begin{pmatrix}1&0&0\\0&1&0\\0&0&-2\end{pmatrix},\qquad \hat{\mathbf{\Lambda}}=\begin{pmatrix}1&0&0\\0&-1&0\\0&0&0\end{pmatrix}.
\end{gathered}
\tag{3.2.4}
$$

For the ground-state structure of dipole moment projections onto the square lattice plane (see Fig. 3.3) we have:

$$
\begin{gathered}
\mathbf{e}_1=(\sin\theta\cos\varphi,\ \sin\theta\sin\varphi,\ \cos\theta),\qquad \mathbf{e}_2=(\sin\theta\cos\varphi,\ -\sin\theta\sin\varphi,\ \cos\theta),\\
\mathbf{e}_3=(-\sin\theta\cos\varphi,\ -\sin\theta\sin\varphi,\ \cos\theta),\mathbf{e}_4=(-\sin\theta\cos\varphi,\ \sin\theta\sin\varphi,\ \cos\theta)
\end{gathered}
\tag{3.2.5}
$$

(where 90° - θ is the dipole inclination to the lattice plane), and the substitution of Eqs. (3.2.4) and (3.2.5) into (3.1.14)-(3.1.16) and (3.1.23) yields the following squared frequencies and integral intensities of spectral lines:

$$
\Omega_1^2=\omega_0^2\left[1+\frac{\chi}{a^3}(1+\kappa)\left(-2D_{\mathrm{F}}\cos^2\theta+D_{\mathrm{A}}\sin^2\theta\right)\right],\qquad A_1\propto\varepsilon_z^2\cos^2\theta,
$$

$$
\Omega_2^2=\omega_0^2\left[1+\frac{\chi}{a^3}(1+\kappa)\left(-2D_{\mathrm{F}}\cos^2\theta+D_{\mathrm{A}}\sin^2\theta\right)-\frac{1}{2}\frac{\chi}{a^3}\left(1-\sqrt{2}\right)D_{\mathrm{F}}\sin^2\theta\right],\ A_2=0,
$$

$$\Omega_3^2 = \omega_0^2\left\{1+\frac{\chi}{a^3}(1+\kappa)\left(-2D_F\cos^2\theta + D_A\sin^2\theta\right)-\frac{1}{2}\frac{\chi}{a^3}\left(1-\sqrt{2}\right)D_F\cos^2\theta,\right.$$
$$\left.+\frac{\chi}{a^3}D_F\left[\cos^2\varphi+\left(2^{-1/2}-1\right)\sin^2\varphi\right]\sin^2\theta\right\},\qquad A_2\propto\varepsilon_x^2\sin^2\theta\cos^2\varphi,$$
$$\Omega_4^2 = \omega_0^2\left\{1+\frac{\chi}{a^3}(1+\kappa)\left(-2D_F\cos^2\theta + D_A\sin^2\theta\right)-\frac{1}{2}\frac{\chi}{a^3}\left(1-\sqrt{2}\right)D_F\cos^2\theta,\right.$$
$$\left.+\frac{\chi}{a^3}D_F\left[\sin^2\varphi+\left(2^{-1/2}-1\right)\cos^2\varphi\right]\sin^2\theta\right\},\qquad A_2\propto\varepsilon_x^2\sin^2\theta\sin^2\varphi, \tag{3.2.6}$$

where $\chi = q^2/m\omega_0^2$ is a molecular vibration polarizability and $\kappa = 2\alpha\mu/qk$ is a dimensionless parameter accounting for static frequency renormalization at the non-zero cubic anharmonicity α. The unitary matrix of transformation to normal coordinates takes the form

$$S_{jl} = \frac{1}{2}\begin{pmatrix} 1 & 1 & 1 & 1 \\ 1 & -1 & 1 & -1 \\ 1 & 1 & -1 & -1 \\ 1 & -1 & -1 & 1 \end{pmatrix}. \tag{3.2.7}$$

Taking into consideration relations (3.2.4), expressions (3.2.6) for absorption lines allowed in infrared spectra ($A_l \neq 0$) are the same as those obtained formerly by an alternative method which involves no sublattice concept.[81] The determination of Davydov splittings as squared frequency differences (3.2.6) results in their independence from the static frequency renormalization. For structures with $\varphi = 0$, which is the case for the system CO/NaCl(100),[28,99] Eqs. (3.2.6) provide good agreement with the observed Davydov splitting[81] (see Sec. 3.3).

As an example, we now apply the relation (3.2.2) to the case of a complex honeycomb lattice,[124] with hexagonal sides equal to a. The basis vectors $\mathbf{A}_1 = \sqrt{3}a(1, 0, 0)$ and $\mathbf{A}_2 = \sqrt{3}a\left(1/2, \sqrt{3}/2, 0\right)$ for the basic triangular sublattice are expressible in terms of the basis vectors $\mathbf{a}_1 = a(0, -1, 0)$ and $\mathbf{a}_2 = a\left(\sqrt{3}/2, -1/2, 0\right)$ for the dense triangular lattice: $\mathbf{A}_1 = 2\mathbf{a}_2 - \mathbf{a}_1$; $\mathbf{A}_1 = \mathbf{a}_2 - 2\mathbf{a}_1$ (see Fig. 3.4 a). Further, $\mathbf{r}_{21} = a\left(\sqrt{3}/2, 1/2, 0\right) = \mathbf{a}_2 - \mathbf{a}_1$ and all the sites of the complex lattice are contained in the site set of the dense Bravais lattice. The basis vectors for the reciprocal lattices are interrelated: $\mathbf{B}_1 = (\mathbf{b}_1 + 2\mathbf{b}_2)/3$ and

$\mathbf{B}_1 = -(2\mathbf{b}_1 + \mathbf{b}_2)/3$, so that the corresponding first Brillouin zones appear as in Fig. 3.4 b and the summation in Eq. (3.2.2) is confined to the vectors $\mathbf{B} = -\mathbf{B}_1, 0, \mathbf{B}_1$.

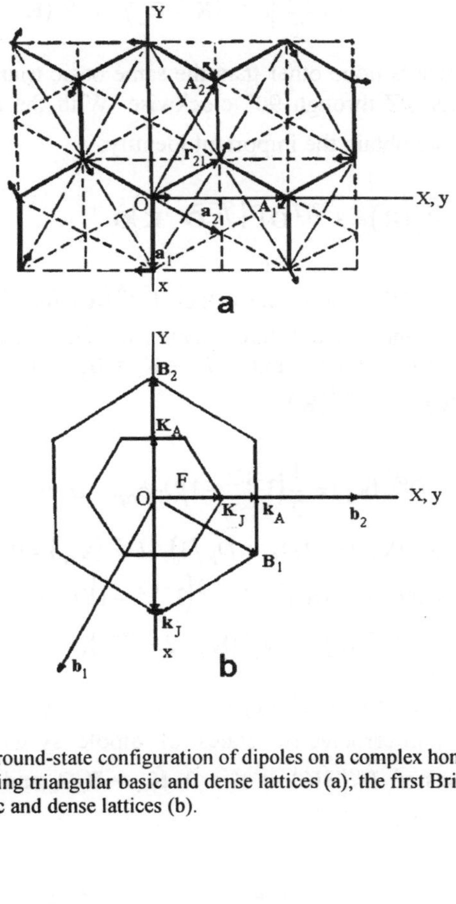

Fig. 3.4. The ground-state configuration of dipoles on a complex honeycomb lattice and the corresponding triangular basic and dense lattices (a); the first Brillouin zones for the reciprocal basic and dense lattices (b).

Thus, we have:

$$\tilde{V}_{11}^{\alpha\beta}(\mathbf{K}) = \tilde{V}_{22}^{\alpha\beta}(\mathbf{K}) = \frac{1}{3}\left[\tilde{V}^{\alpha\beta}(\mathbf{K}) + \tilde{V}^{\alpha\beta}(\mathbf{K}+\mathbf{B}_2) + \tilde{V}^{\alpha\beta}(\mathbf{K}-\mathbf{B}_2)\right], \qquad (3.2.8)$$

$$\tilde{V}_{21}^{\alpha\beta}(\mathbf{K}) = \left[\tilde{V}_{12}^{\alpha\beta}(\mathbf{K})\right]^{*} = \frac{1}{3}\left\{\tilde{V}^{\alpha\beta}(\mathbf{K}) - \frac{1}{2}\left[\tilde{V}^{\alpha\beta}(\mathbf{K}+\mathbf{B}_2) + \tilde{V}^{\alpha\beta}(\mathbf{K}-\mathbf{B}_2)\right]\right.$$
$$\left. + i\frac{\sqrt{3}}{2}\left[\tilde{V}^{\alpha\beta}(\mathbf{K}+\mathbf{B}_2) - \tilde{V}^{\alpha\beta}(\mathbf{K}-\mathbf{B}_2)\right]\right\}. \qquad (3.2.9)$$

The dense Bravais lattice is none other than the same basic triangular lattice merely rotated around the axis OZ through 90° clockwise. With the appropriate rotation matrix $\hat{\mathbf{O}}$ introduced, we obtain the important identity:

$$\hat{\tilde{V}}_{jj}(\mathbf{K}) = 3^{-3/2}\hat{\mathbf{O}}\hat{\tilde{V}}\left(\sqrt{3}\,\hat{\mathbf{O}}^{-1}\mathbf{K}\right)\hat{\mathbf{O}}^{-1} \qquad (3.2.10)$$

which, along with Eq. (3.2.8), allows the values $\tilde{V}^{\alpha\beta}(\mathbf{K})$ for different points of the first Brillouin zone to be interrelated. Specifically, for the symmetric points $\mathbf{K} = 0$, $\mathbf{k}_A = \mathbf{b}_2/2$, $\mathbf{k}_J = -\mathbf{B}_2$ (see Fig. 3.4 b), and at $\mathbf{K} = \mathbf{K}_J = \mathbf{b}_2/3$, we derive the following expressions for $D^{\alpha\beta}(\mathbf{K}) = a^3\tilde{V}^{\alpha\beta}(\mathbf{K})$:

$$\begin{aligned}
&D^{\alpha\beta}(0) = D_F\delta_{\alpha\beta},\ D^{\alpha\beta}(\mathbf{k}_J) = \frac{1}{2}\left(3^{-1/2}-1\right)D_F\delta_{\alpha\beta},\ \alpha,\beta = x,y,\\
&D^{xx}(\mathbf{k}_A) = D_A,\ D^{yy}(\mathbf{k}_A) = -(D_A + D_F/3),\ D^{xy}(\mathbf{k}_A) = 0,\\
&D^{xx}(\mathbf{K}_J) = -4.453809,\ D^{yy}(\mathbf{K}_J) = 3^{-3/2}\left(3^{-1/2}-1\right)D_F - D^{xx}(\mathbf{K}_J),\\
&D^{xy}(\mathbf{K}_J) = 0, \qquad D^{zz}(\mathbf{K}) = -D^{xx}(\mathbf{K}) - D^{yy}(\mathbf{K}),
\end{aligned} \qquad (3.2.11)$$

where the parameters $D_F = -5.517088$ and $D_A = -4.094910$ determine the energies of the ferroelectric and antiferroelectric states of dipole systems on a triangular lattice.[56] Now the values of $D_{21}^{\alpha\beta}(\mathbf{K})$ at $\mathbf{K} = 0$, $\mathbf{K}_A = \mathbf{B}_2/2$, and $\mathbf{K}_J = \mathbf{b}_2/3$ (see Fig. 3.4 b) are derivable from Eq. (3.2.9):

$$\begin{aligned}
&D_{21}^{\alpha\beta}(0) = \frac{1}{2}\left(1-3^{-3/2}\right)D_F\delta_{\alpha\beta},\ D_{21}^{XX}(\mathbf{K}_A) = \frac{1}{2}\left[\left(1-3^{-3/2}\right)D_A + \frac{1}{3}D_F\right]\exp\left(-i\frac{\pi}{3}\right),\\
&D_{21}^{YY}(\mathbf{K}_A) = -\frac{1}{2}\left[\left(1+3^{-3/2}\right)D_A + 3^{-5/2}D_F\right]\exp\left(-i\frac{\pi}{3}\right),\ D_{21}^{XY}(\mathbf{K}_A) = 0,\\
&D_{21}^{XX}(\mathbf{K}_J) = -D_{21}^{YY}(\mathbf{K}_J) = -i\,D_{21}^{XY}(\mathbf{K}_J)\\
&\quad = \frac{1}{4}3^{-3/2}\left(3^{-1/2}-1\right)D_F - \frac{1}{2}D^{xx}(\mathbf{K}_J),
\end{aligned} \qquad (3.2.12)$$

$$D_{21}^{ZZ}(\mathbf{K}) = -D_{21}^{XX}(\mathbf{K}) - D_{21}^{YY}(\mathbf{K}).$$

The eigenvalues of tensor $D_{jj'}^{\alpha\beta}(\mathbf{K})$ at the symmetric points of the first Brillouin zone that are determined by formula (3.1.7) take the form:

$$\begin{aligned}
&D_1^1(0) = D_1^2(0) = -\frac{1}{2}D_1^3(0) = \frac{1}{2}\left(1+3^{-3/2}\right)D_{\mathrm{F}} = -3.289426,\\
&D_2^1(0) = D_2^2(0) = -\frac{1}{2}D_2^3(0) = \frac{1}{2}\left(3^{-3/2}-1\right)D_{\mathrm{F}} = 1.165898,\\
&D_1^1(\mathbf{K}_{\mathrm{A}}) = \frac{1}{2}\left[\frac{1}{3}D_{\mathrm{F}} + \left(1+3^{-1/2}\right)D_{\mathrm{A}}\right] = -4.149068,\\
&D_1^2(\mathbf{K}_{\mathrm{A}}) = -\frac{1}{2}\left[3^{-5/2}D_{\mathrm{F}} - \left(1-3^{-3/2}\right)D_{\mathrm{A}}\right] = -1.476461,\\
&D_2^1(\mathbf{K}_{\mathrm{A}}) = -\frac{1}{2}\left[\frac{1}{3}D_{\mathrm{F}} + \left(1-3^{-3/2}\right)D_{\mathrm{A}}\right] = 2.572937,\\
&D_2^2(\mathbf{K}_{\mathrm{A}}) = -\frac{1}{2}\left[3^{-3/2}D_{\mathrm{F}} + \left(1+3^{-3/2}\right)D_{\mathrm{A}}\right] = 3.760436,\\
&D_1^3(\mathbf{K}_{\mathrm{A}}) = -D_1^1(\mathbf{K}_{\mathrm{A}}) - D_2^2(\mathbf{K}_{\mathrm{A}}),\quad D_2^3(\mathbf{K}_{\mathrm{A}}) = -D_1^2(\mathbf{K}_{\mathrm{A}}) - D_2^1(\mathbf{K}_{\mathrm{A}}),\\
&D_1^1(\mathbf{K}_{\mathrm{J}}) = D^{xx}(\mathbf{K}_{\mathrm{J}}) = -4.453809,\quad D_2^2(\mathbf{K}_{\mathrm{J}}) = D^{yy}(\mathbf{K}_{\mathrm{J}}) = 4.902564,\\
&D_1^2(\mathbf{K}_{\mathrm{J}}) = D_2^1(\mathbf{K}_{\mathrm{J}}) = -\frac{1}{2}D_1^3(\mathbf{K}_{\mathrm{J}}) = -\frac{1}{2}D_2^3(\mathbf{K}_{\mathrm{J}}) = \frac{1}{2}\left[3^{-3/2}\left(3^{-1/2}-1\right)\right]D_{\mathrm{F}} = 0.224377.
\end{aligned} \tag{3.2.13}$$

The dependences $D_p^{\nu}(\mathbf{K})$ calculated along symmetric directions within the first Brillouin zone are given in Fig. 3.5. The minimum value, $D_1^1(\mathbf{K}_{\mathrm{J}})$, determines the energy $H_0 = (1/2)N(\mu^2/a^3)D_1^1(\mathbf{K}_{\mathrm{J}})$ of the one-parameter-degenerate (in angle φ) ground state with the following configuration of dipole moments:

$$\begin{aligned}
\mathbf{e}_{\mathbf{R}1} &= \left[\cos(\mathbf{K}_{\mathrm{J}}\cdot\mathbf{R}+\varphi),\ \sin(\mathbf{K}_{\mathrm{J}}\cdot\mathbf{R}+\varphi),\ 0\right]\\
\mathbf{e}_{\mathbf{R}2} &= \left[-\cos(\mathbf{K}_{\mathrm{J}}\cdot\mathbf{R}+\varphi),\ \sin(\mathbf{K}_{\mathrm{J}}\cdot\mathbf{R}+\varphi),\ 0\right]
\end{aligned} \tag{3.2.14}$$

(see Fig. 3.4 a). Interestingly, the same ground-state configuration can be arrived at by considering dipole-dipole interactions in terms of the simplest nearest-neighbour approximation (at $D_1^1(\mathbf{K}_{\mathrm{J}})$= -9/2)[125] whereas for a triangular lattice of dipoles, short- and long-range interaction models lead to ground-state structures which are qualitatively different.[56]

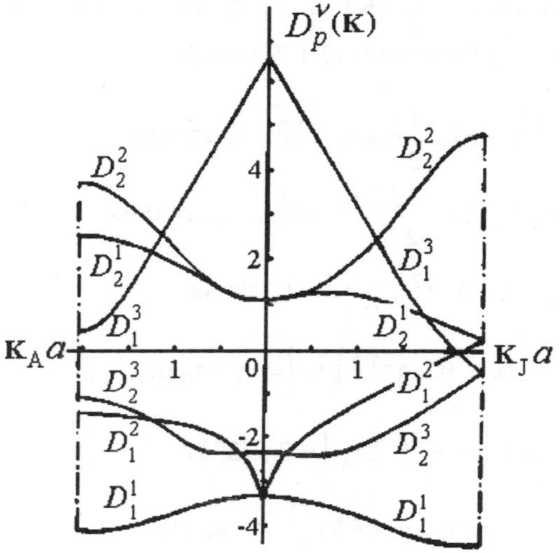

Fig. 3.5. Fourier component eigenvalues $D_p^\nu(\mathbf{K})$ for the dipole-dipole interaction tensor on a honeycomb lattice plotted versus the wave vector **K** determined in the first Brillouin zone of the basic sublattice.

Let us analyse the effect caused by the thermodynamic fluctuations of order parameter on the ground state of a dipole honeycomb lattice. Two factors determine the situation just as they do in the case of a square lattice.[54,56,81] These are the quadratic asymptotics of the minimum-energy branch $D_1^1(\mathbf{K}_J)$ of the eigenvalues in the vicinity of the ground state with $\mathbf{K} = \mathbf{K}_J$ and one-parametric ground-state degeneracy in the angle variable (see (2.2.22)). As was shown previously for a square lattice,[70] these factors lead the system concerned to exist in a long-range order phase at low temperatures.

The curves $D_p^\nu(\mathbf{K})$ shown in Fig. 3.5 correspond to six branches of vibrations of charges which can shift relative to the honeycomb lattice sites. As a lattice of this sort actually exists on the basic graphite face, for which a great body of experimental adsorption evidence has been collected,[126,127] it is appropriate to present the calculated radial vibrational frequencies for adsorbed polar molecules that should be observable in the infrared spectra. Adsorption potentials can force the dipole moments of adsorbed molecules out of the lattice plane[28] so that the X and Y components of the vectors in Eq. (3.2.14) are multiplied by $\sin^2\theta$ and the Z components amount to $\cos^2\theta$. Based on the calculations performed previously for a square lattice of adsorption centres[81] we are led to

$$\Omega_Z^2 = \omega_0^2\left\{1 + \frac{\chi}{a^3}(1+\kappa)\left[D_1^3(0)\cos^2\theta + D_1^1(\mathbf{K}_J)\sin^2\theta\right]\right\},$$

$$\Omega_X^2 = \Omega_Y^2 = \omega_0^2\left\{1 + \frac{\chi}{a^3}\kappa\left[D_1^3(0)\cos^2\theta + D_1^1(\mathbf{K}_J)\sin^2\theta\right]\right. \tag{3.2.15}$$

$$\left. + \frac{\chi}{a^3}D_1^3(\mathbf{K}_J)\cos^2\theta + \frac{1}{2}\frac{\chi}{a^3}\left[D_1^1(0) + D_1^2(\mathbf{K}_J)\right]\sin^2\theta\right\}.$$

Here the indices X, Y, Z indicate vibration polarizations, other designations being the same as in expressions (3.2.6). Finally, the Davydov splitting of spectral lines obtained as

$$\Delta\Omega_{\mathrm{Dav}} \approx \frac{1}{2\omega_0}\left(\Omega_Z^2 - \Omega_X^2\right) =$$

$$\frac{\chi\omega_0}{2a^3}\left\{\left[D_1^3(0) - D_1^3(\mathbf{K}_J)\right]\cos^2\theta - \left[\frac{1}{2}\left(D_1^1(0) + D_1^2(\mathbf{K}_J)\right) - D_1^1(\mathbf{K}_J)\right]\sin^2\theta\right\} \tag{3.2.16}$$

is governed by the values of $D_p^\nu(\mathbf{K})$ (see Eq. (3.2.13)) at the symmetric points of the first Brillouin zone and originates from the lateral interactions solely of the dynamic dipole moments of adsorbed molecules.

3.3. Chain orientational structures on planar lattices

As seen from Chapter 2, adsorbed molecules often form monolayers with chain orientational structures in which the chains with identically oriented molecules alternate (Fig. 2.4). Consider the Davydov splitting of vibrational spectral lines in such systems. Let molecular orientations be specified by the angles θ_j and φ_j in the spherical coordinate system with the z-axis perpendicular to the lattice plane:

$$\mathbf{e}_j = \left(\sin\theta_j\cos\varphi_j,\ \sin\theta_j\sin\varphi_j,\ \cos\theta_j\right), \qquad j = 1, 2\,. \tag{3.3.1}$$

where two possible values of the index j correspond to two chains of identically oriented molecules. Fig. 2.4 depicts a particular case of such herringbone structure with $\theta_2 = \theta_1 \equiv \theta$ and $\varphi_2 = \pi - \varphi_1 \equiv \varphi$ which is of most interest to us. This situation implies that $\Phi_{11} = \Phi_{22}$ and the solution of Eq. (3.1.16) appears especially simple:

$$\begin{matrix} \Omega_A = \Phi_{11} - \Phi_{12} \\ \Omega_S = \Phi_{11} + \Phi_{12} \end{matrix} \quad S = \frac{1}{\sqrt{2}}\begin{pmatrix} 1 & 1 \\ -1 & 1 \end{pmatrix}, \tag{3.3.2}$$

where the indices l = A and l = S corresponding to the first and the second columns of the matrix S denote collectivized antisymmetric and symmetric vibrations. Introduce polarization vectors $\mathbf{P}_l$ for these vibrations:

$$\mathbf{P}_l = \sum_{j=1}^{2} S_{jl}\mathbf{e}_j = \begin{cases} \sqrt{2}(\sin\theta\cos\varphi,\, 0,\, 0), & l = \mathrm{A}, \\ \sqrt{2}(0,\, \sin\theta\sin\varphi,\, \cos\theta), & l = \mathrm{S}. \end{cases} \tag{3.3.3}$$

As easily seen, the vector $\mathbf{P}_A$ is always parallel to the surface plane and the surface-parallel component of the vector $\mathbf{P}_S$ vanishes for the structures having collinear projections of molecular axes onto the surface plane ($\varphi = 0$) (see Fig. 3.6).

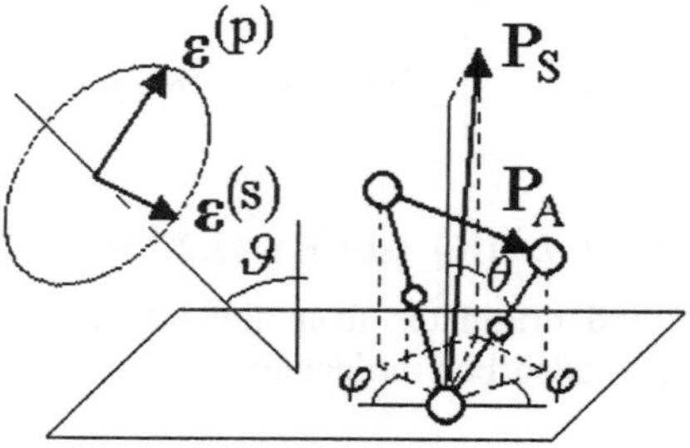

Fig. 3.6. Polarization vectors $\mathbf{P}_l$ for symmetric (l = S) and antisymmetric (l = A) vibrations.

Relation (3.1.25) for the integral intensity of the lth spectral line for s- and p-polarized radiation is conveniently expressed in terms of polarization vectors (3.3.3):

$$A_l^{(p,s)} = \frac{\pi^2 N\chi\omega_0^2}{c_0 F\cos\vartheta}\left|\left(\boldsymbol{\varepsilon}^{(p,s)}\cdot\mathbf{P}_l\right)\right|^2, \quad l = \mathrm{S, A}\ . \tag{3.3.4}$$

Substituting Eqs. (3.1.26) and (3.3.3) in the above expression, we obtain:

$$\frac{A_{\mathrm{S}}^{(s)}}{A_{\mathrm{A}}^{(s)}} = \tan^2\varphi, \qquad \frac{A_{\mathrm{S}}^{(p)}}{A_{\mathrm{A}}^{(p)}} = \frac{\tan^2\theta\sin^2\varphi + 2\lambda\tan^2\vartheta}{\tan^2\theta\cos^2\varphi}, \qquad \lambda = \left(\frac{1+r^{(p)}}{1-r^{(p)}}\right)^2. \quad (3.3.5)$$

For quasinormal orientations, $\tan\theta << 1$, and it is straightforward to derive from Eq. (3.3.5) that $A_{\mathrm{S}}^{(p)}/A_{\mathrm{A}}^{(p)} >> 1$. This inference is illustrated well by monolayer transmittance spectra of the adsorbate $^{12}C^{16}O/NaCl(100)$ (see Fig. 2.8): The spectral line of the symmetric collective vibration of CO molecules in the *p*-polarized spectrum is much more intensive than that of the antisymmetric collective vibration. Owing to $\varphi = 0$ in this system, the *s*-polarized spectrum lacks the line of the symmetric collective vibration. For comparison, the adsorbate $^{12}C^{16}O_2/NaCl(100)$ does not belong to systems with quasinormal orientations and its spectrum is accordingly devoid of the above-considered features. Eqs. (3.3.5) are very convenient in geometrical determination of herringbone orientational structures. The azimuthal angle φ is easy to find from the integral intensity ratio for s-polarized spectral lines (see the first relation in Eq. (3.3.5)); then the polar angle θ is given by the integral intensity ratio for p-polarized spectral lines (see the second relation in Eq. (3.3.5)). It is clear that the first and the second relations in Eq. (3.3.5) coincide for planar orientational structures ($\theta = 90^0$).

Values of the force matrix components Φ_{11} and Φ_{12} are specified by the corresponding Fourier components of the dipole-dipole interaction tensors $\widetilde{V}_{jj'}^{\alpha\beta}(\mathbf{K})$ at $\mathbf{K} = 0$ (see (3.1.14) or (3.1.24)). Computational techniques to provide these quantities for simple and complex Bravais lattices are respectively discussed in Secs. 2.1 and 3.2. As the structures considered in this section represent the chains of identically oriented molecules, it is expedient to take advantage of the chain representation of dipole-dipole interactions (see Sec. 2.1). The quantity Φ_{11} is contributed by interactions within chains (see relation (2.2.10) generalized to the dipole moment deviations from the chain axis) and between the chains spaced at the interval *b* (see Fig. 2.4). The magnitude of Φ_{12} is dictated by the interactions between the chains with differently oriented dipole moments (separated by the minimum distance *b*/2). Interchain interactions are constituted by two components falling off with distance by the power and exponential laws. The former arises only with the proviso that the dipole moment orientations are noncollinear with respect to chain axes and can be described in the continual approximation. To this end, a chain of identically oriented molecules is regarded as two uniformly charged closely-spaced parallel threads with the charge densities q/a and $-q/a$, where the effective charge q and the dipole length l are related by the dipole moment value, μ

$= ql$. For the orientational structure presented in Fig. 2.4, the displacements of the two threads relative to one another, if projected onto the axes z and y, amount to $l\cos\theta$ and $l\sin\theta\sin\varphi$ (by multiplying these projections by q/a, the corresponding projections of the dipole moment density are obtainable). The electric field induced by two threads at the distance y reckoned along the axis y, is expressed as:

$$\mathbf{E}(y) = \frac{2\mu}{ay^2}(0,\, \sin\theta\sin\varphi,\, \cos\theta). \tag{3.3.6}$$

The interchain interaction of dipole moments collinear to the chain axes decays exponentially with the interchain distance (see Eq. (2.2.11)) and has no contribution descending by the power law y^{-2} in Eq. (3.3.6). Summation over lattice sites and interchain distances involves the Riemann zeta-function

$$\zeta(p) = \sum_{n=1}^{\infty} n^{-p} \tag{3.3.7}$$

with $p = 3$ and 2 ($\zeta(3) \approx 1.202$, $\zeta(2) \approx 1.645$). In the ground state determination for dipole systems, exponentially small terms are of prime importance: it is because of them that the interchain interaction energy depends on chain longitudinal displacements. On the other hand, exponentially small terms contribute only about 10% to the interchain interaction energy and can thus be neglected in the energy estimation. By virtue of this simplification, the interchain interactions are describable in a very simple analytical form. Indeed, the explicit expressions for the force matrix components Φ_{jj} ($j = 1, 2$) and Φ_{12}

$$\begin{aligned}\Phi_{jj} &\approx \omega_j^2\left\{1 + \frac{2\zeta(3)\chi_j}{a^3}\left[\cos^2\theta_j + \left(1 - 3\cos^2\varphi_j\right)\sin^2\theta_j\right]\right.\\ &\qquad\left. + \frac{4\zeta(2)\chi_j}{ab^2}\left[\cos^2\theta_j - \sin^2\theta_j\sin^2\varphi_j\right]\right\},\\ \Phi_{12} &\approx \frac{12\zeta(2)\omega_1\omega_2\sqrt{\chi_1\chi_2}}{ab^2}\left[\cos\theta_1\cos\theta_2 - \sin\theta_1\sin\theta_2\sin\varphi_1\sin\varphi_2\right]\end{aligned} \tag{3.3.8}$$

depend only on the parameters ω_j and χ_j for the molecular vibrations concerned, molecular orientations in a couple of chains, the separation a between the neighboring lattice sites in the chains, and the interchain distance $b/2$.[128] The approximation exploited involves no dependence of force matrix components on the mutual displacement of lattice sites in neighboring chains, which substantially extends the applicability of relations (3.3.8).

Let us define the Davydov splitting of spectral lines, $\Delta\Omega_{Dav}$, as a frequency difference between the symmetric and antisymmmetric collective vibrations. For a herringbone structure with $\theta_2 = \theta_1$ and $\varphi_2 = \pi - \varphi_1$ (as well as for $\omega_1 = \omega_2 \equiv \omega_0$ and $\chi_1 = \chi_2 \equiv \chi$), substituting Eq. (3.3.8) into (3.3.2) and the use of the inequality $\Delta\Omega_{Dav} << \omega_0$ provide:

$$\Delta\Omega_{Dav} \equiv \Omega_S - \Omega_A \approx \frac{\Phi_{12}}{\omega_0} \approx 12\zeta(2)\omega_0 \frac{\chi}{ab^2}\left[\cos^2\theta - \sin^2\theta \sin^2\varphi\right]. \qquad (3.3.9)$$

Now we pass on the analysis of the relations derived focusing on several particular cases of importance which enable us of correlating the calculated values with the available experimental data. CO molecules adsorbed on the (100) face of a NaCl crystal reside at the sites of a square lattice ($a = b/2 = 3.988$ Å); at sufficiently low temperatures ($T < 25$ K), they have inclined orientations ($\theta = 25°$) with alternating dipole moment projections onto the axes of the neighboring chains ($\varphi_1 = 0$, $\varphi_2 = 180°$).[28] For this system, the Davydov splitting of vibration spectral lines is determined as:

$$\Delta\Omega_{Dav} \approx 3.935\left(\chi/a^3\right)\cos^2\theta\,. \qquad (3.3.10)$$

The accurate allowance made for dipole-dipole interactions leads to:[81]

$$\Delta\Omega_{Dav} = 3.995\left(\chi/a^3\right)\left(1 - 1.058\sin^2\theta\right), \qquad (3.3.11)$$

which relation only slightly differs from the approximate one (see Eq. (3.3.10)). Substituting $\omega_0 = 2155$ cm^{-1} and $\chi = 0.057$ Å^3 (vibration polarizability of CO molecules in gaseous phase) into Eq. (3.3.10) yields $\Delta\Omega_{Dav} \approx 7.8$ cm^{-1}, in good accord with the observed quantity $\Delta\Omega_{Dav} \approx 6$ cm^{-1}.[28]

CO_2 molecules adsorbed on the same surface form a rhombic lattice ($a = b/2 = 3.988$ Å) with the orientational parameters $\theta = 65°$, $\varphi_1 = 49.5°$, $\varphi_2 = 130.5°$.[13-26,123] With such values of the angular variables, expression (3.3.9) goes negative; with the corresponding vibrational parameters entered ($\omega_0 = 2349$ cm^{-1} and $\chi = 0.48$ Å^3 in gaseous phase[129]), the Davydov splitting becomes about -26 cm^{-1}. At the same time, a materially smaller absolute value of the quantity concerned was recorded in experiments: $\Delta\Omega_{Dav} \approx -9.2$ cm^{-1}.[123] To apprehend the discrepancy, one should take into account that the Davydov splitting scale is dictated by the difference of trigonometric functions (see the bracketed expression in formula (3.3.9)) which is very sensitive to the relation between surface-parallel and surface-normal components of dipole-dipole interactions. These components are differently screened by a substrate. To estimate the screening effect, we employ a simplistic model based on image force of point dipoles. Consider two point dipoles A and B, with the dipole moments $\boldsymbol{\mu}_A = \boldsymbol{\mu}_A^{\|} + \boldsymbol{\mu}_A^z$ and $\boldsymbol{\mu}_B = \boldsymbol{\mu}_B^{\|} + \boldsymbol{\mu}_B^z$, the heights above the surface z_A and z_B, and the mutual displacement along the surface **r**. The overall interaction between them is contributed by the interactions with point dipole moments of their images, $\boldsymbol{\mu}_{A'} = \boldsymbol{\mu}_A^{\|} - \boldsymbol{\mu}_A^z$ and $\boldsymbol{\mu}_{B'} = \boldsymbol{\mu}_B^{\|} - \boldsymbol{\mu}_B^z$, induced in the substrate bulk at the distances z_A and z_B from the interface between vacuum and a medium with the dielectric constant ε:

$$U_{\text{Dip}} = D^{\alpha\beta}(\mathbf{r})\mu_A^\alpha \mu_B^\beta + \frac{1}{2}\frac{1-\varepsilon}{1+\varepsilon}\left[D^{\alpha\beta}(\mathbf{r}+2\mathbf{z}_A)\mu_{A'}^\alpha \mu_B^\beta + D^{\alpha\beta}(\mathbf{r}-2\mathbf{z}_B)\mu_A^\alpha \mu_{B'}^\beta\right], \quad \alpha,\ \beta = x,\ y,\ z. \tag{3.3.12}$$

At $\varepsilon \to 1$ or at z_A, $z_B >> r$, the second summand approaches zero, as might be expected. In the other limiting case, at $z_A = z_B \equiv z << r$, the following power series results (accurate to terms $(z/r)^2$):

$$U_{\text{Dip}} = \frac{2\varepsilon}{1+\varepsilon}\left[1 - 9\frac{\varepsilon-1}{\varepsilon}\left(\frac{z}{r}\right)^2\right]\frac{\mu_A^z \mu_B^z}{r^3} + \frac{2}{1+\varepsilon}\left[\delta_{\alpha\beta} - 3\frac{r^\alpha r^\beta}{r^2} + 3(\varepsilon-1)\left(\delta_{\alpha\beta} - 5\frac{r^\alpha r^\beta}{r^2}\right)\left(\frac{z}{r}\right)^2\right]\frac{\mu_A^\alpha \mu_B^\beta}{r^3}, \quad \alpha,\beta = x,y. \tag{3.3.13}$$

As is obvious from the above relation, the screening effect of the substrate on interactions causes, in the limiting case $z \to 0$, the change in the surface-normal components by a factor of $2\varepsilon/(1+\varepsilon)$ and in the surface-parallel components by a factor of $2/(1+\varepsilon)$, i.e., the renormalization ratio is equal to ε. As the substrate dielectric constant increases, the interaction of surface-parallel components monotonically decays to zero, whereas the interaction of surface-normal components is enhanced reaching the maximum, viz. the interaction with a double dipole moment, for a metal (at $\varepsilon \to \infty$).[130] Analogous renormalization factors are also of significance in the treatment of island-like particles on a dielectric substrate.[131] With such effects included, formula (3.3.9) can be rewritten as:

$$\Delta\Omega_{\rm Dav} \approx \frac{24\zeta(2)}{1+\varepsilon}\frac{\chi}{ab^2}\omega_0\left[\varepsilon\cos^2\theta - \sin^2\theta\sin^2\varphi\right], \tag{3.3.14}$$

which results in the understated absolute magnitude of the Davydov splitting, $\Delta\Omega_{\rm Dav} \approx -3.3\ \mathrm{cm}^{-1}$, for the system CO_2/NaCl(100) ($\varepsilon = n^2 \approx 2.31$ with $n \approx 1.52$ denoting the refractive index of NaCl in the spectral range of stretching CO_2–vibrations). Analyzing the corrections of the order $(z/r)^2$ in Eq. (3.3.14), one can notice that they somewhat attenuate the screening effect, so that the result tends to the experimental value, $\Delta\Omega_{\rm Dav} \approx -9.2\ \mathrm{cm}^{-1}$.

As another illustration, we consider the two-dimensional aggregates of benzothiazolocarbocyanine dyes (BTCC) adsorbed on the (111) face of AgBr crystal. The face of interest presents a triangular lattice constituted by Ag^+ ions, with a lattice constant a equal to 4.082 Å. The separation between neighboring molecular planes within the stacks (regarded as chains), $r_0 = a \cdot \sin\varphi$ (with $\varphi = 60°$), matches the separation of Ag^+ ion rows so that $r_0 \approx 3.536$ Å.[132] The neighboring stacks can have either the same ($\varphi_2 = \varphi_1$)[132] or different ($\varphi_2 = \pi - \varphi_1$)[133] molecular orientations; the latter case presents a herringbone structure shown in Fig. 2.4. The equidistant arrangement of molecular stacks corresponds to the accommodation of each BTCC molecule on six Ag^+ ions so that $L = b/(2\sin\varphi) \approx 6 \times 4.082$ Å ≈ 24.492 Å. The frequency shift caused by the interaction between the neighboring stacks is specified by the half-scale of the Davydov splitting (as defined in Eq. (3.3.9) at $\theta = 90°$) and electronic polarizability $\chi = 2\mu^2/(\hbar\omega_0)$ (here μ is a matrix element of the transition dipole moment):[128]

$$\omega_{12} = \frac{1}{2}\Delta\Omega_{\rm Dav} \approx -\frac{3\zeta(2)\mu^2\sin\varphi}{\hbar r_0 L^2}. \tag{3.3.15}$$

At $\varphi_2 = \varphi_1$, the polarization $\mathbf{P}_A$ vanishes and the only spectral line of symmetric collective excitations is observed. At $\varphi_2 = \pi - \varphi_1$, the integral intensity ratio for the s- and p-polarized spectral lines (the former corresponding to the lower frequency) which arise from the splitting is given by the ratio of the corresponding polarizations (see Eqs. (3.3.4) and (3.3.5) at $\theta = 90°$):

$$|\mathbf{P}_S|^2 / |\mathbf{P}_A|^2 = \tan^2\varphi\ . \tag{3.3.16}$$

The dipole moments of the first electronic transitions in polymethine dye molecules can be derived in the context of the long-polymethine-chain approximation:[128,134,135]

$$\mu = -\left(2/\pi^2\right) e r^{\|}_{\text{C-C}} (n+1+\lambda). \tag{3.3.17}$$

Here $r^{\|}_{\text{C-C}} \approx 1.247$ Å is the projection of a chain C-C bond onto the long molecular axis; n is the number of C atoms in the chain; λ is the effective length of the dye end-groups which is easily calculable from their topological characteristics.[134-136] Given $n = 3$ and $\lambda = 4.58$ for BTCC molecule, we come to $\mu = 10.4$ D which estimate agrees perfectly both with the value observed in the experiment ($\mu_{\text{exp}} = 10.5$ D) and with that calculated by the Hückel LCAO-MO method ($\mu_{\text{calc}} = 10.2$ D).[132]

For two stacks having the same molecular orientations, the numerical factor $3\zeta(2)$ in formula (3.3.15) should be substituted by 2; as a result, with $\varphi = 60°$ and the parameters indicated above, the frequency shift yielded in the point-dipole approximation is $\omega_{12} = -445$ cm^{-1}. On the other hand, the computer simulation of two parallel stacks, each consisting of ten BTCC molecules, gave the frequency shift arising from the interstack Coulomb interaction: $\omega_{12}^{\text{calc}} = -446$ cm^{-1}.[132] The excellent accord between these two values suggests that the long-chain approximation is quite efficient in the calculations of individual molecular characteristics of polymethine dyes and the point-dipole approximation is likewise justified when applied to the treatment of interchain interactions in aggregates.

The intrachain dipole-dipole interactions of BTCC molecules are responsible for the frequency shift contributing to the first of Eqs. (3.3.8) as $2\zeta(3)\mu^2(1 - 3\cos^2\varphi)/(\hbar a^3)$, with its sign changing at $\varphi = 54°44'$. However, the straightforward calculations of the Coulomb interactions show this boundary angle to be equal to 30°.[132] Thus, as far as intrastack interactions of dye molecules are concerned, the point-dipole approximation introduces large errors into the treatment.

For the herringbone BTCC aggregate with $\varphi = 30°$ adsorbed on AgBr(111) surface, the spectral line of the first electronic transition was observed[133] to be Davydov-split into two components, at 645 and 600 nm (i.e. $\Delta\Omega_{\text{Dav}}^{\text{exp}} \approx 1160$ cm^{-1}), the latter component being thrice as intensive as the former just as relation (3.3.16) dictates. With the appropriate values of the parameters mentioned above, Eq. (3.3.15) leads to: $\Delta\Omega_{\text{Dav}} = 2\omega_{12} \approx 1270$ cm^{-1} which value exceeds the experimental one by no more than 10%.

To sum up, the analytical model proposed presents a rare instance when a rather simple treatment enables not only qualitative but also quantitative relationships between structure and optical properties to be revealed.

We now turn to a more complex instance of the Davydov-split spectral lines corresponding to the bending vibrations of CO_2 molecules in a monolayer adsorbed on the NaCl(100) surface. A molecule CO_2 in gaseous phase exhibits two degenerate bending vibrations in the plane perpendicular to the long molecular axis.

On adsorption, this axis is parallel to the unit vector specified by Eq. (3.3.1) and the orientations of the bending vibration appear as follows:

$$\mathbf{e}_{js} = \left(-\sin\varphi_j,\ \cos\varphi_j,\ 0\right),\ \mathbf{e}_{jp} = \left(-\cos\theta_j\cos\varphi_j,\ -\cos\theta_j\sin\varphi_j,\ \sin\theta_j\right) \\ j = 1,\ 2 \qquad (3.3.18)$$

(Eqs. (3.3.1) and (3.3.18) define a triad of mutually orthogonal vectors). Evidently, the adsorption potential induced by the substrate removes the degeneration of the bending vibrations, oriented parallel to ($\mathbf{e}_{js}$) and inclined at the angle of 90° - θ to the surface ($\mathbf{e}_{jp}$). The singleton frequencies were obtained by means of isotopic mixture experiments, that is, recording the IR spectra of mixtures of $^{12}C^{16}O_2$ and $^{13}C^{16}O_2$ as a function of the concentrations. In the limit of infinite dilution of $^{12}C^{16}O_2$, the IR spectrum showed a single band at ω_0 = 2349.0 cm^{-1} in the range of the asymmetric stretching vibration, whereas in the range of the bending vibration the extrapolation yielded ω_s =660.0 cm^{-1} and ω_p =655.5 cm^{-1}.[137] Since the singleton frequency ω_0 of the asymmetric stretching vibration on the surface is much higher than the corresponding values ω_s and ω_p of the two bending modes, the solutions of equation (3.1.16), with allowance made for several molecular modes, become approximately separated for $v = 0$ and $v = s, p$, respectively:

$$\sum_{j'v'} \Phi_{jv,j'v'} S_{j'v',l} = \Omega_l^2 S_{jv,l}\ . \qquad (3.3.19)$$

Eigenvalues and eigenvectors for the asymmetric stretching vibration of herringbone structure with $\theta_2 = \theta_1 \equiv \theta$ and $\varphi_2 = \pi - \varphi_1 \equiv \varphi$ are described by relations (3.3.2) and (3.3.8). In order to determine the eigenvalues Ω_l^2 and eigenvectors $S_{jv,l}$ of the two bending modes s and p we have to consider the following 4×4 matrix:

$$\widetilde{\Phi}_{sp} = \begin{pmatrix} \Phi_{1s,1s} & \Phi_{1s,2s} & \Phi_{1s,1p} & \Phi_{1s,2p} \\ \Phi_{1s,2s} & \Phi_{1s,1s} & -\Phi_{1s,2p} & -\Phi_{1s,1p} \\ \Phi_{1s,1p} & -\Phi_{1s,2p} & \Phi_{1p,1p} & \Phi_{1p,2p} \\ \Phi_{1s,2p} & -\Phi_{1s,1p} & \Phi_{1p,2p} & \Phi_{1p,1p} \end{pmatrix}. \qquad (3.3.20)$$

It is convenient to introduce the two pair indices $\lambda = \pm 1$ and $\mu = \pm 1$ instead of l. We can now specify the eigenvalues and eigenvectors by means of the following analytical expressions:[123]

$$\Omega^2_{\lambda\mu} = \frac{1}{2}\left[\Phi_{1s,1s} + \Phi_{1p,1p} + \lambda\left(\Phi_{1s,2s} - \Phi_{1p,2p}\right) + \mu\Lambda_\lambda\right],$$
$$S_{(s/p),\lambda\mu} = \sqrt{\left|\frac{\Phi_{1s,1p} - \lambda\Phi_{1s,2p}}{2\Lambda_\lambda D_{\lambda\mu}}\right|}\begin{pmatrix} D_{\lambda\mu} \\ \lambda D_{\lambda\mu} \\ 1 \\ -\lambda \end{pmatrix} \tag{3.3.21}$$

with

$$\Lambda_\lambda = \sqrt{\left[\Phi_{1s,1s} - \Phi_{1p,1p} + \lambda\left(\Phi_{1s,2s} - \Phi_{1p,2p}\right)\right]^2 + 4\left[\Phi_{1s,1p} - \lambda\Phi_{1s,2p}\right]^2},$$
$$D_{\lambda\mu} = \frac{\Phi_{1s,1s} - \Phi_{1p,1p} + \lambda(\Phi_{1s,2s} - \Phi_{1p,2p}) + \mu\Lambda_\lambda}{2(\Phi_{1s,1p} - \lambda\Phi_{1s,2p})}. \tag{3.3.22}$$

Substitution of $S_{jv,l}$ in Eq. (3.1.25) by the corresponding vector components given by Eq. (3.3.21) leads to the integrated absorption of the spectral line with indices λ and μ:

$$A_{\lambda\mu} = \frac{\pi^2 N}{2c_0 F \cos\vartheta}\left|\frac{\Phi_{1s,1p} - \lambda\Phi_{1s,2p}}{\Lambda_\lambda D_{\lambda\mu}}\right|$$
$$\times\left|\boldsymbol{\varepsilon}\cdot\left[\omega_s\sqrt{\chi_s}D_{\lambda\mu}\left(\mathbf{e}_{1s} + \lambda\mathbf{e}_{2s}\right) + \omega_p\sqrt{\chi_p}\left(\mathbf{e}_{1p} - \lambda\mathbf{e}_{2p}\right)\right]\right|^2. \tag{3.3.23}$$

The peak frequencies, integral intensities, and $A_l^{(s)}/A_l^{(p)}$ ratios for stretching and bending vibrations of the CO_2/NaCl(100) system are listed in Table 3.1 together with the measured values.[123] The calculations involve the above-presented parameters of the system as well as the vibrational polarizabilities in gaseous phase, χ_0 = 0.48 Å^3 and $\chi_s = \chi_p$ = 0.24 Å^3.[129] The comparison of calculated and experimental data shows a good agreement both for the polarizations (except for the absolute value of the $A_l^{(s)}/A_l^{(p)}$ ratio for row 5 in Table 3.1) and for the frequencies, especially in the case of bending vibrations. Some discrepancies for stretching vibration frequencies may be attributable to the already discussed screening effect of a substrate.

Table 3.1. Calculated IR peak frequencies, integral intensities and $A_l^{(s)}/A_l^{(p)}$ ratios for stretching (A and S) and bending (λ, μ) vibrations in CO_2/NaCl(100) monolayer using the parameters given in the text. The measured values are written in parentheses.[123]

l (λ, μ)	Ω_h (cm^{-1})	$A_l^{(s)}$	$A_l^{(p)}$	$A_l^{(s)}/A_l^{(p)}$
A	2354.7 (2349.06)	0.074 (0.079)	0.044 (0.046)	1.695 (1.72)
S	2332.7 (2339.89)	0.102 (0.105)	0.133 (0.133)	0.768 (0.79)
(1, 1)	660.5 (659.26)	0.0028 (0.0045)	0.0016 (0.0025)	1.742 (1.80)
(1, -1)	653.5 (652.78)	0.0030 (0.0050)	0.0017 (0.0035)	1.742 (1.43)
(-1, 1)	663.8 (663.24)	0.0010 (0.0045)	0.0134 (0.0080)	0.073 (0.56)
(-1,-1)	654.6 (655.79)	0.0037 (0.0035)	0.0021 (0.0025)	1.738 (1.40)

Chapter 4

Temperature dependences of spectral line shifts and widths

Treating vibrational excitations in lattice systems of adsorbed molecules in terms of bound harmonic oscillators (as presented in Chapter III and also in Appendix 1) provides only a general notion of basic spectroscopic characteristics of an adsorbate, viz. spectral line frequencies and integral intensities. This approach, however, fails to account for line shapes and manipulates spectral lines as shapeless infinitely narrow and infinitely high images described by the Dirac δ-functions. In simplest cases, the shape of symmetric spectral lines can be characterized by their maximum positions and full width at half maximum (FWHM). These parameters are very sensitive to various perturbations and changes in temperature and can therefore provide additional evidence on the state of an adsorbate and its binding to a surface.

Among the accepted mechanisms governing spectral line shapes are inhomogeneous broadening inherent in disordered systems and homogeneous broadening arising from the interaction of a localized or collectivized excitation with the environment. For vibrational excitations in an adsorbate, the environment is represented by the phonon thermostat of a substrate. Harmonic coupling between the vibrational displacements of atoms in the adsorbate and in the substrate can lead to a finite broadening of adsorbate spectral lines only with the proviso that the corresponding frequency falls within the quasicontinuous spectrum of the substrate. This condition is met for the low-frequency vibrations of adsorbed molecules which involve rotational or translational (along the surface) degrees of freedom. To account for the spectral line broadening for high-frequency stretching vibrations of adsorbed molecules, allowance should be made for the anharmonic coupling of these vibrations with the low-frequency modes of the same molecules (the latter have finite lifetimes due to the harmonic coupling with the quasicontinuous spectrum of the substrate). The situation becomes even more involved as a consequence of the collectivization of the high-frequency and low-frequency vibrations due to lateral intermolecular interactions.

This wide range of questions is to be elucidated in the present chapter. The bulk of attention is given to the effects induced by the collectivization of adsorbate vibrational modes whose low-frequency components are coupled to the phonon thermostat of the substrate. This coupling gives rise to the resonant nature of low-frequency collective excitations of adsorbed molecules (see Sec. 4.1). A mechanism underlying the occurrence of resonance (quasilocal) vibrations is most readily comprehended using the example of a system of harmonic oscillators in which the states of a certain oscillator, due to its coupling with all the others, have finite lifetimes. These problems are considered in detail in Appendix 1.

Adsorption potential and lateral interactions convert rotational orientational states intrinsic in molecules in gaseous phase to hindered-rotation states typical of adsorbed molecules. Small vibrations around the minima of the hindered-rotation potential can in the rough be regarded as harmonic only providing sufficiently high reorientation barriers. Then it is admissible to treat orientational excitations as quasiparticles obeying the Bose-Einstein statistics and to apply the corresponding retarded Green's function technique. Actually, reorientation barriers are not, however, too high and a comparably small number of orientational vibration quanta fits in their heights. In view of this, thermally activated reorientations and subbarrier tunneling should be taken into account. Sec. 4.2 presents the corresponding technique for computing spectroscopic characteristics; it includes thermally activated reorientations as a stochastic process of a system's departure from subbarrier equidistant energy levels. Here we introduce the Markov approximation for retarded Green's functions which underlies an exact expression for a spectral line shape in the case of biquadratic anharmonic coupling between two oscillators in the phonon field of a substrate (the exchange dephasing model). Also, the perturbation theory for the Pauli equation is presented which refers to the subbarrier group of states. The results obtained are adjusted to the limiting case of weak interaction between the subsystem concerned and the phonon field of the thermostat so as to apply them to a description of one-sided temperature broadening for H-bond molecular complexes. For a better understanding of this section, Appendix 2 provides the classical picture of thermally activated reorientations and quantum tunnel relaxation of orientational states in the phonon field of a substrate.

In Sec. 4.3, we generalize the exchange dephasing model to various cases of anharmonic coupling between high-frequency and low-frequency modes, and to the case of collectivized excitations in adsorbate. For the system H(D)/C(111), the high-frequency and low-frequency modes differ in frequencies by a factor of about 2. Therefore, cubic anharmonic coupling of these modes contributes materially to spectral function characteristics of high-frequency vibrations. The systems with sufficiently strong lateral interactions of high-frequency molecular modes are distinguished by a strong dependence of spectral function parameters on the molecular axis inclination to the surface plane. Lateral interactions of low-frequency molecular modes also exert influence on the spectral function of high-frequency vibrations; an illustrative example is provided by the analysis of temperature-dependent shifts and widths of the Davydov-split spectral lines for 2x1 phase of the monolayer CO/NaCl(100). All relevant mathematical derivations (based on double-time Green's functions in the representation of Matsubara's frequency space) can be found in Appendix 3.

4.1. Resonant nature of low-frequency collective excitations of adsorbed molecules

Here it is our intention to show that for a system constituted by substrate phonons and laterally interacting low-frequency adsorbate vibrations which are harmonically coupled with the substrate, the states can be subclassified into independent groups by the wave vector **K** referring to the first Brillouin zone of the adsorbate lattice.[138] As the phonon state density of a substrate many-fold exceeds the vibrational mode density of an adsorbate, for each adsorption mode there is a quasicontinuous phonon spectrum in every group of states determined by **K** (see Fig. 4.1). Consequently, we can regard the low-frequency collectivized mode of the adsorbate, $\omega_l(\mathbf{K})$, as a resonance vibration with the renormalized frequency ω_{K} and the reciprocal lifetime η_{K}.

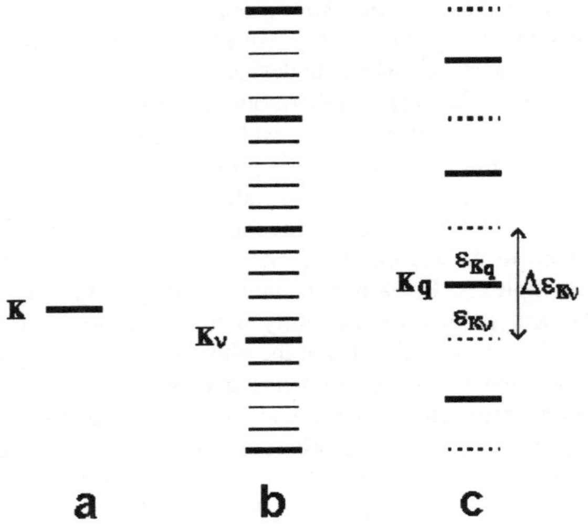

Fig. 4.1. Coupling of the adsorbate low-frequency mode with substrate phonons: **K** level of the adsorbate (a); initial quasicontinuous phonon spectrum of the substrate not perturbed by the adsorbate, bold lines designating the levels which correspond to the specified wave vector **K** (b); level shifts in the **K** subsystem caused by the coupling of the adsorbate **K** mode and substrate phonons (c).

Represent the Hamiltonian for low-frequency modes of the adsorbate and substrate as a sum of three terms:

$$H_\ell = H_\ell^{(\mathrm{mol})} + H_S + H_{\mathrm{int}}, \tag{4.1.1}$$

the first of them accounting for laterally interacting adsorbate modes:

$$\begin{aligned} H_\ell^{(\mathrm{mol})} &= \sum_{\mathbf{R}} \frac{p_\ell^2(\mathbf{R})}{2m_\ell} + \frac{1}{2}\sum_{\mathbf{R},\mathbf{R}'}\left[m_\ell\omega_\ell^2\delta_{\mathbf{R},\mathbf{R}'} + \Phi_{\ell,lat}(\mathbf{R}-\mathbf{R}')\right]u_\ell(\mathbf{R})\,u_\ell(\mathbf{R}') \\ &= \sum_{\mathbf{K}} \hbar\omega_\ell(\mathbf{K})\left(b_\ell^+(\mathbf{K})b_\ell(\mathbf{K}) + 1/2\right). \end{aligned} \tag{4.1.2}$$

Here $u_l(\mathbf{R})$ and $p_l(\mathbf{R})$ denote the shift and generalized momentum for the molecular vibration of the low frequency ω_φ and reduced mass m_φ at the $\mathbf{R}th$ site of the adsorbate lattice; $b_l^+(\mathbf{K})$ and $b_l(\mathbf{K})$ are creation and annihilation operators for the collectivized mode of the adsorbate that is characterized by the squared frequency $\omega_l^2(\mathbf{K}) = \omega_l^2 + \widetilde{\Phi}_{l,\mathrm{lat}}(\mathbf{K})/m_l$, with $\widetilde{\Phi}_{l,\mathrm{lat}}(\mathbf{K})$ representing the Fourier component of the force constant function $\Phi_{l,\mathrm{lat}}(\mathbf{R})$. Shifts $u_l(\mathbf{R})$ for all molecules are assumed to be oriented in the same arbitrary direction specified by the unit vector $\mathbf{e}_l$; they are related to the corresponding normal coordinates, $\widetilde{u}_\ell(\mathbf{K})$, and secondary quantization operators:

$$u_\ell(\mathbf{R}) = (m_\ell N_0)^{-1/2}\sum_{\mathbf{K}} \widetilde{u}_\ell(\mathbf{K})e^{iK\cdot\mathbf{R}}, \quad \widetilde{u}_\ell(\mathbf{K}) = \left(\frac{\hbar}{2\omega_\ell(\mathbf{K})}\right)^{1/2}\left(b_\ell(\mathbf{K}) + b_\ell^+(-\mathbf{K})\right), \tag{4.1.3}$$

where N_0 is a number of adsorbate lattice sites in the main area.

The second summand in the right side of Eq. (4.1.1) describes substrate phonons:

$$H_S = \sum_{\mathbf{k}_{||},\sigma} \hbar\omega_S(\mathbf{k}_{||},\sigma)\left(b_{\mathbf{k}_{||},\sigma}^+ b_{\mathbf{k}_{||},\sigma} + 1/2\right). \tag{4.1.4}$$

In the above relation, quantum states of phonons are characterized by the surface-parallel wave vector $\mathbf{k}_{||}$, whereas the rest of quantum numbers are indicated by σ; the latter account for the polarization of a quasi-particle and its motion in the surface-normal direction, and also implicitly reflect the arrangement of atoms in the crystal unit cell. A convenient representation like this allows us to immediately take advantage of the translational symmetry of the system in the surface-parallel direction so as to define an arbitrary Cartesian projection (onto the α axis) for the shift of a substrate atom with the mass M residing at the $\mathbf{R}th$ site of the adsorbate lattice:

$$u_S^{\alpha}(\mathbf{R}) = (MN_0)^{-1/2} \sum_{\mathbf{k}_{||},\sigma} S_{\mathbf{k}_{||},\sigma}^{\alpha} e^{i\mathbf{k}_{||}\cdot\mathbf{R}} \widetilde{u}_S(\mathbf{k}_{||},\sigma),$$
$$\widetilde{u}_S(\mathbf{k}_{||},\sigma) = \left(\frac{\hbar}{2\omega_S(\mathbf{k}_{||},\sigma)}\right)^{1/2} \left(b_{\mathbf{k}_{||},\sigma} + b_{-\mathbf{k}_{||},\sigma}^{+}\right) \tag{4.1.5}$$

(here $S_{\mathbf{k}_{||},\sigma}^{\alpha}$ are unitary matrices affording the switch to normal coordinates $\widetilde{u}_S(\mathbf{k}_{||},\sigma)$).

In this representation, the third summand in the right side of the Eq. (4.1.1) which describes a harmonic coupling between low-frequency vibrations of the adsorbate and substrate assumes a substantially simplified form:

$$H_{\text{int}} = \sum_{\mathbf{R},\alpha,\beta} \Phi_{\text{int}}^{\alpha\beta} e_{\ell}^{\alpha} u_{\ell}(\mathbf{R}) u_S^{\beta}(\mathbf{R})$$
$$= (Mm_{\ell})^{-1/2} \sum_{\mathbf{K},\mathbf{k}_{||},\sigma,\alpha,\beta} e_{\ell}^{\alpha} \Phi_{\text{int}}^{\alpha\beta} S_{\mathbf{k}_{||},\sigma}^{\beta} \widetilde{u}_{\ell}(\mathbf{K}) \widetilde{u}_S(\mathbf{k}_{||},\sigma) N_0^{-1} \sum_{\mathbf{R}} e^{i(\mathbf{K}+\mathbf{k}_{||})\cdot\mathbf{R}}. \tag{4.1.6}$$

The first Brillouin zone for vectors $\mathbf{k}_{||}$, being determined by the crystal unit cell, it can be larger than that for the adsorbate lattice, and hence the sum over $\mathbf{R}$ entering into the Eq. (4.1.6) is found as

$$\sum_{\mathbf{R}} e^{i(\mathbf{k}_{||}+\mathbf{K})\cdot\mathbf{R}} = N_0 \sum_{\mathbf{B}} \delta_{\mathbf{k}_{||}+\mathbf{K},\mathbf{B}}, \tag{4.1.7}$$

where summation over $\mathbf{B}=n_1\mathbf{B}_1+n_2\mathbf{B}_2$ is performed over all inequivalent integer linear combinations of the vectors $\mathbf{B}_1$ and $\mathbf{B}_2$ of the adsorbate reciprocal lattice. This allows the classification of the wave vector set, $\mathbf{k}_{||}$, into vector subsets $\mathbf{B}$-$\mathbf{K}$, and expression (4.1.6) can hence be rewritten as follows:

$$H_{\text{int}} = \frac{1}{2} \sum_{\mathbf{K},\nu} \left[\chi_{\mathbf{K},\nu} \widetilde{u}_{\ell}(\mathbf{K}) \widetilde{u}_S^{*}(\mathbf{K},\nu) + \chi_{\mathbf{K},\nu}^{*} \widetilde{u}_{\ell}^{*}(\mathbf{K}) \widetilde{u}_S(\mathbf{K},\nu)\right], \tag{4.1.8}$$

where

$$\chi_{\mathbf{K},\nu} = (Mm_\ell)^{-1/2} \sum_{\alpha,\beta} e_\ell^\alpha \Phi_{\text{int}}^{\alpha\beta} S_{\mathbf{B}-\mathbf{K},\sigma}^\beta , \tag{4.1.9}$$

and the new quantum number ν serves instead of the quantum number couple σ and **B**.

The summation over $\mathbf{k}_\parallel$ and σ in formula (4.1.4) can also be represented as a sum over **K** and ν variables. Then Eqs. (4.1.1), (4.1.2), (4.1.4), and (4.1.8) lead to an inference of fundamental importance: the Hamiltonian H_ℓ for low-frequency modes of the adsorbate and substrate represents a sum of Hamiltonians $H_\ell(\mathbf{K})$, i.e. it is diagonal in terms of the wave vector **K** for the adsorbate. On this basis, an independent transformation to new normal coordinates $x_{\mathbf{K}q}$ in each diagonal block $H_\ell(\mathbf{K})$ is possible:

$$\tilde{u}_\ell(\mathbf{K}) = \sum_q C_{\mathbf{K},\mathbf{K}q} x_{\mathbf{K}q}, \quad \tilde{u}_S(\mathbf{K},\nu) = \sum_q C_{\mathbf{K}\nu,\mathbf{K}q} x_{\mathbf{K}q} . \tag{4.1.10}$$

Unitary matrices $C_{\mathbf{K}q}$ and $C_{\mathbf{K}\nu,\mathbf{K}q}$ are diagonal in **K**, the former transforming from adsorbate to low-frequency system normal coordinates and satisfying the equations:

$$\left[\omega_{\mathbf{K}q}^2 - \omega_\ell^2(\mathbf{K}) - g_{\mathbf{K}}\left(\omega_{\mathbf{K}q}^2\right)\right] C_{\mathbf{K},\mathbf{K}q} = 0, \quad \left[1 - g'_{\mathbf{K}}(\varepsilon)\big|_{\varepsilon=\omega_{\mathbf{K}q}^2}\right] \left|C_{\mathbf{K},\mathbf{K}q}\right|^2 = 1, \tag{4.1.11}$$

where

$$g_{\mathbf{K}}(\varepsilon) = \sum_\nu \frac{\left|\chi_{\mathbf{K},\nu}\right|^2}{\varepsilon - \omega_S^2(\mathbf{K},\nu)} . \tag{4.1.12}$$

Let us now invoke the conception of systems with a quasicontinuous spectrum developed by Lifshits.[139] Since energy levels $\varepsilon_{\mathbf{K}q} = \omega_{\mathbf{K}q}{}^2$ for a perturbed system and levels $\varepsilon_{\mathbf{K}\nu} = \omega_S{}^2(\mathbf{K},\nu)$ for the corresponding unperturbed one alternate (see Fig. 4.1), the position of an arbitrary level $\varepsilon_{\mathbf{K}q}$ on the interval $(\varepsilon_{\mathbf{K}\nu}, \varepsilon_{\mathbf{K}\nu} + \Delta\varepsilon_{\mathbf{K}\nu})$ can be reckoned from the level $\varepsilon_{\mathbf{K}\nu}$ and characterized by the parameter $\zeta_{\mathbf{K}q}$ varying from 0 to 1: $\varepsilon_{\mathbf{K}q} = \varepsilon_{\mathbf{K}\nu} + \zeta_{\mathbf{K}q}\Delta\varepsilon_{\mathbf{K}\nu}$. The function $g_{\mathbf{K}}(\varepsilon_{\mathbf{K}q})$ and its derivative are hence expressible as:[135]

$$g_{\mathbf{K}}\left(\varepsilon_{\mathbf{K}q}\right) = \tilde{\mathrm{P}}_{\mathbf{K}}(\varepsilon_{\mathbf{K}\nu}) + \pi\tilde{\eta}_{\mathbf{K}}(\varepsilon_{\mathbf{K}\nu}) \cot \pi\zeta_{\mathbf{K}q} , \tag{4.1.13}$$

$$g'_{\mathbf{K}}\left(\varepsilon_{\mathbf{K}q}\right)=1+\frac{1}{\Delta\varepsilon_{\mathbf{K}q}}\frac{\partial g_{\mathbf{K}}\left(\varepsilon_{\mathbf{K}\nu}\right)}{\partial\zeta_{\mathbf{K}q}},$$
$$\frac{\partial g_{\mathbf{K}}\left(\varepsilon_{\mathbf{K}\nu}\right)}{\partial\zeta_{\mathbf{K}q}}=-\frac{\left[\varepsilon_{\mathbf{K}\nu}-\omega_{\varphi}^{2}(\mathbf{K})-\widetilde{\mathrm{P}}_{\mathbf{K}}\left(\varepsilon_{\mathbf{K}\nu}\right)\right]^{2}+\pi^{2}\widetilde{\eta}_{\mathbf{K}}^{2}\left(\varepsilon_{\mathbf{K}\nu}\right)}{\widetilde{\eta}_{\mathbf{K}}\left(\varepsilon_{\mathbf{K}\nu}\right)}. \tag{4.1.14}$$

Detailed derivation of formula (4.1.13) with an emphasis on the origin of the term $\cot\pi\zeta_{\mathbf{K}q}$ is given in Appendix 2 (see Eq. (A2.83)). Formulae (4.1.14) are obtained by substituting characteristics of the quasicontinuous spectrum into the first of Eqs. (4.1.11), with regard to Eq. (4.1.13) and the relationship between perturbed and unperturbed interlevel gaps, $\Delta\varepsilon_{\mathbf{K}q}{}^{-1}=[\Delta\varepsilon_{\mathbf{K}\nu}{}^{-1}-d\zeta_{\mathbf{K}\nu}/d\varepsilon_{\mathbf{K}\nu}]^{-1}$. The quantities $\widetilde{\eta}_{\mathbf{K}}(\varepsilon)$ and $\widetilde{\mathrm{P}}_{\mathbf{K}}(\varepsilon)$ are calculable by the equations:

$$\widetilde{\eta}_{\mathbf{K}}(\varepsilon)=-\frac{1}{\pi}\operatorname{Im}g_{\mathbf{K}}(\varepsilon+i0)=\sum_{\nu}\left|\chi_{\mathbf{K}\nu}\right|^{2}\delta\left(\varepsilon-\varepsilon_{\mathbf{K}\nu}\right),$$
$$\widetilde{\mathrm{P}}_{\mathbf{K}}(\varepsilon)=\operatorname{Re}g_{\mathbf{K}}(\varepsilon+i0)=\int_{0}^{\infty}\frac{\widetilde{\eta}_{\mathbf{K}}(\widetilde{\varepsilon})d\widetilde{\varepsilon}}{\varepsilon-\widetilde{\varepsilon}}, \tag{4.1.15}$$

where Re and Im denote real and imaginary parts of a complex function, respectively. Fig. 4.1 schematically depicts the alternating levels $\varepsilon_{\mathbf{K}\nu}$ and $\varepsilon_{\mathbf{K}q}$ which belong to the group of **K**-dependent states.

Incorporating expression (4.1.13) into the equation $g_{\mathbf{K}}(\varepsilon_{\mathbf{K}q})=\varepsilon_{\mathbf{K}q}-\omega_{l}^{2}(\mathbf{K})$ which follows from relation (4.1.11) and calculating the derivative in Eq. (4.1.11) in view of Eq. (4.1.14) we arrive at:[140]

$$\left|C_{\mathbf{K},\mathbf{K}q}\right|^{2}=\left[-\frac{\partial g_{\mathbf{K}}\left(\varepsilon_{\mathbf{K}q}\right)}{\partial\zeta_{\mathbf{K}q}}\right]^{-1}\Delta\varepsilon_{\mathbf{K}q}$$
$$=\frac{\widetilde{\eta}_{\mathbf{K}}\left(\varepsilon_{\mathbf{K}q}\right)\Delta\varepsilon_{\mathbf{K}q}}{\left[\varepsilon_{\mathbf{K}q}-\omega_{l}^{2}(\mathbf{K})-\widetilde{P}_{\mathbf{K}}\left(\varepsilon_{\mathbf{K}q}\right)\right]^{2}+\pi^{2}\widetilde{\eta}_{\mathbf{K}}^{2}\left(\varepsilon_{\mathbf{K}q}\right)}\approx\Re\left(\omega_{\mathbf{K}q}-\omega_{\mathbf{K}};\eta_{\mathbf{K}}\right)\Delta\omega_{\mathbf{K}q}, \tag{4.1.16}$$

where

$$\Re(\omega;\eta) \equiv \frac{1}{2\pi}\frac{\eta}{\omega^2 + (\eta/2)^2} \tag{4.1.17}$$

is the resonance function, $\omega_{\mathbf{K}}$ is the resonance frequency obeying the equation

$$\omega_{\mathbf{K}}^2 - \omega_l^2(\mathbf{K}) - \tilde{P}_{\mathbf{K}}\left(\omega_{\mathbf{K}}^2\right) = 0, \tag{4.1.18}$$

and the parameter

$$\eta_{\mathbf{K}} = \frac{\pi}{\omega_{\mathbf{K}}}\tilde{\eta}_{\mathbf{K}}\left(\omega_{\mathbf{K}}^2\right), \tag{4.1.19}$$

if multiplied by 2π, accounts for the full width of the resonance function. As the Fourier transform of the resonance function (4.1.17) takes on the form

$$\Re(t;\eta) = \frac{1}{2\pi}\int_{-\infty}^{\infty}\Re(\omega;\eta)e^{-i\omega t}d\omega = \frac{1}{2\pi}e^{-\eta|t|/2}, \tag{4.1.20}$$

the lifetime of the resonance mode with the wave vector **K** is found as $\tau_{\mathbf{K}} = 2/\eta_{\mathbf{K}}$.

The quasimode approximation is justified only in the case of $\eta_{\mathbf{K}} \ll \omega_{\mathbf{K}}$, i.e., for long lifetimes of resonance modes. Then resonance function (4.1.17) is narrow enough and all slowly varying functions of $\omega_{\mathbf{K}q}$ can be substituted by their values at the point $\omega_{\mathbf{K}q} = \omega_{\mathbf{K}}$. It has been just this procedure that has led us to the last approximate equality in relation (4.1.16). Importantly, the summation over q in expressions containing $|C_{\mathbf{K},\mathbf{K}q}|^2$ corresponds, in view of relation (4.1.16), to the switch to integral sums involving the resonance function which, in turn, can be replaced by integrals. Thus, if $F(\omega)$ is an arbitrary slowly varying function of ω, then

$$\sum_q F\left(\omega_{\mathbf{K}q}\right)\left|C_{\mathbf{K},\mathbf{K}q}\right|^2 = \int_{-\infty}^{\infty}F(\omega)\Re(\omega-\omega_{\mathbf{K}};\eta_{\mathbf{K}})d\omega = F(\omega_{\mathbf{K}}) \tag{4.1.21}$$

and the result is independent of $\eta_{\mathbf{K}}$. In the special case that $F \equiv 1$, we revert to the corresponding normalization condition.

Explicit expressions for the dispersion laws $\omega_{\mathbf{K}}$ and resonance width $\eta_{\mathbf{K}}$ can be deduced for a model system, with the substrate simulated by a semi-infinite elastic continuum and the adsorbate overlayer presented as an array of point masses connected to the surface by harmonic springs.[141,142] These expressions are strongly coverage-dependent and predict relaxation rates in good quantitative agreement with available experiments. A remarkable feature of this treatment is that no energy transfer from the adsorbate to the surface is possible at $K > \omega_l/c_T$ (c_T is transverse sound velocity) and hence $\eta_{\mathbf{K}} = 0$ in this region.[143] Averaging over all the wave vectors **K** of the first Brillouin zone furnishes the resonance width of a single adsorbate (see formulae (4.2.25) and (A1.109)).

4.2. Markov approximation for Green's functions of molecular subsystems in the condensed phase

Practically, any experimental study of an arbitrary system reduces to measuring the response of some physical characteristic A of the system to a probing external action which perturbs, in the general case, another characteristic B. The required response is then described by a conveniently calculable retarded Green's function (GF) that contains sufficiently complete information about the states of the system:[144]

$$G(t) = -i\theta(t)\left\langle\left[\hat{A}(t), \hat{B}(0)\right]\right\rangle, \quad \hat{A}(t) = \exp\left(\frac{i}{\hbar}\hat{H}t\right)\hat{A}\exp\left(-\frac{i}{\hbar}\hat{H}t\right),$$
$$\langle\ldots\rangle = \mathrm{Sp}(\hat{\rho}\ \ldots), \quad \hat{\rho} = \exp\left(-\hat{H}/T\right)/\mathrm{Sp}\left[\exp\left(-\hat{H}/T\right)\right], \tag{4.2.1}$$

where $\hat{A}$ and $\hat{B}$ are the operators of the physical quantities A and B, and $\hat{\rho}$ is the equilibrium statistical operator of the system determined by its Hamiltonian $\hat{H}$ and the absolute temperature T (in energy units). In many cases one is interested in the properties of a molecular subsystem with a Hamiltonian $\hat{H}^{(\mathrm{mol})}$ which is coupled through an interaction $\hat{H}_{\mathrm{int}} \equiv \hat{V}$ with the remaining large part (surface reservoir) of the total system described by the Hamiltonian $\hat{H}_S$. We have then

$$\hat{H} = \hat{H}^{(\mathrm{mol})} + \hat{H}_S + \hat{V} \tag{4.2.2}$$

and the operators $\hat{A}$ and $\hat{B}$ occur only in the Hamiltonian $\hat{H}^{(\mathrm{mol})}$. This fact by itself does not lead to any simplification of Eq. (4.2.1), since the operator $\hat{V}$ containing the variables of the subsystem and of the reservoir does not commute with $\hat{H}^{(\mathrm{mol})}$ or $\hat{H}_S$.

On the other hand, there exist well-developed methods for calculating states of subsystems using the Markov approximation for the reduced density matrix (statistical operator) of the subsystem $\hat{\rho}^{(\mathrm{mol})} = \mathrm{Tr}\hat{\rho}$ (Tr indicates the trace over the variables of the reservoir).[145] The average value of a physical quantity $A(t)$ will now be determined by the trace of the product of operator matrices for the subsystem only:

$$A(t) = \mathrm{Sp}\left(\hat{\rho}^{(\mathrm{mol})}\hat{A}\right) \tag{4.2.3}$$

and the response to an external action described by the time-dependent operator $\hat{H}_B(t)$ will be included in the operator $\hat{\rho}^{(\mathrm{mol})}(t)$ satisfying the equation

$$\begin{aligned}
&\frac{\partial}{\partial t}\hat{\rho}^{(\mathrm{mol})}(t) + \frac{i}{\hbar}\left[\hat{H}^{(\mathrm{mol})} + \left\langle \hat{V} \right\rangle_S + \hat{H}_B(t), \hat{\rho}^{(\mathrm{mol})}(t)\right] \\
&= \frac{1}{\hbar^2}\int_{-\infty}^{0} d\tau\, \mathrm{Tr}\left[\Delta\hat{V}, \left[\exp\left[\frac{i}{\hbar}\left(\hat{H}^{(\mathrm{mol})} + \hat{H}_S\right)\tau\right]\Delta\hat{V},\right.\right. \\
&\qquad \left.\left. \times \exp\left[-\frac{i}{\hbar}\left(\hat{H}^{(\mathrm{mol})} + \hat{H}_S\right)\tau\right], \hat{\rho}^{(\mathrm{mol})}(t)\hat{\rho}_S\right]\right], \\
&\Delta\hat{V} = \hat{V} - \left\langle \hat{V} \right\rangle_S, \quad \left\langle \hat{V} \right\rangle_S = \mathrm{Tr}\left(\hat{\rho}_S\hat{V}\right),
\end{aligned} \tag{4.2.4}$$

where $\hat{\rho}_S$ is the equilibrium reservoir operator (the expression with a nonvanishing operator $\left\langle \hat{V} \right\rangle_S$ is derived in detail in Ref. 61).

We find the linear response of a subsystem in contact with a reservoir to an external perturbation involving some variable $\hat{B}$ of the subsystem and depending on the time through a function $F(t)$, so that the corresponding perturbation operator can be written in the form

$$\hat{H}_B(t) = -\hat{B}F(t). \tag{4.2.5}$$

We substitute (4.2.5) into Eq. (4.2.4) and look for a solution of the latter in the form

$$\hat{\rho}^{(\mathrm{mol})}(t) = \hat{\rho}_0 + \delta\hat{\rho}(t),$$

where $\hat{\rho}_0$ is the equilibrium statistical operator of the subsystem. The matrix elements of the linear response $\delta\hat{\rho}(t)$ to the perturbation $\hat{H}_B(t)$ represented in the basis of the eigenstates of the Hamiltonian $\hat{H}^{(\mathrm{mol})}$ ($\delta\rho_{qq'}(t)=\langle q|\delta\hat{\rho}(t)|q'\rangle$, $\hat{H}^{(\mathrm{mol})}|q\rangle = \varepsilon_q|q\rangle$) will satisfy the following equation:

$$\frac{\partial}{\partial t}\delta\rho_{qq'}(t)+\sum_{\tilde{q}\tilde{q}'}\delta\rho_{\tilde{q}\tilde{q}'}(t)\left(i\Omega_{\tilde{q}\tilde{q}'qq'}+\Gamma_{\tilde{q}\tilde{q}'qq'}\right)=\frac{i}{\hbar}B_{qq'}\left(\rho_{q'}-\rho_q\right)F(t). \tag{4.2.6}$$

Here ρ_q is the diagonal matrix element of the equilibrium statistical operator $\hat{\rho}_0$ of the subsystem and the matrices Ω and Γ with four indices are given as follows (see, e.g., Ref. 61 where these quantities are obtained taking into account the nonvanishing operator $\langle\hat{V}\rangle_S$ and the frequency shifts arising from principal-value integrals which are usually neglected in other sources):

$$\Omega_{\tilde{q}\tilde{q}'qq'}=\Omega_{qq'}\delta_{\tilde{q}q}\delta_{\tilde{q}'q'}-\frac{1}{\hbar}\left(\langle\hat{V}\rangle_{\tilde{q}'q'}\delta_{\tilde{q}q}-\langle\hat{V}\rangle_{\tilde{q}q}\delta_{\tilde{q}'q'}\right)$$
$$-\frac{1}{2\pi\hbar^2}\int_{-\infty}^{\infty}d\omega\left\{\sum_r\left[\frac{F_{qrr\tilde{q}}(\omega)}{\Omega_{r\tilde{q}}+\omega}\delta_{\tilde{q}'q'}+\frac{F_{\tilde{q}rrq'}(\omega)}{\Omega_{\tilde{q}'r}-\omega}\delta_{\tilde{q}q}\right]\right.$$
$$\left.-\frac{F_{\tilde{q}'q'q\tilde{q}}(\omega)}{\Omega_{q\tilde{q}}+\omega}-\frac{F_{\tilde{q}'q'q\tilde{q}}(\omega)}{\Omega_{\tilde{q}'q'}-\omega}\right\},$$

$$\Gamma_{\tilde{q}\tilde{q}'qq'}=\frac{1}{2\hbar^2}\left\{\sum_r\left[F_{qrr\tilde{q}}\left(\Omega_{\tilde{q}r}\right)\delta_{\tilde{q}'q'}+F_{\tilde{q}rrq'}\left(\Omega_{\tilde{q}'r}\right)\delta_{\tilde{q}q}\right]\right. \tag{4.2.7}$$
$$\left.-F_{\tilde{q}'q'q\tilde{q}}\left(\Omega_{\tilde{q}q}\right)-F_{\tilde{q}'q'q\tilde{q}}\left(\Omega_{\tilde{q}'q'}\right)\right\},$$

$$F_{\tilde{q}\tilde{q}'qq'}(\omega)=\int_{-\infty}^{\infty}d\tau e^{i\omega\tau}\left\langle\Delta\hat{V}_{\tilde{q}\tilde{q}'}(\tau)\Delta\hat{V}_{qq'}\right\rangle_S,$$

$$\Delta\hat{V}_{qq'}(\tau)=\exp\left(\frac{i}{\hbar}\hat{H}_S\tau\right)\Delta\hat{V}_{qq'}\exp\left(-\frac{i}{\hbar}\hat{H}_S\tau\right),$$

$$\Omega_{qq'}=\frac{1}{\hbar}\left(\varepsilon_q-\varepsilon_{q'}\right).$$

We introduce the GF (with four indices) for the left-hand side of Eq. (4.2.6):[146]

$$\frac{\partial}{\partial t} g_{qq'\tilde{q}\tilde{q}'}(t) + \sum_{q_1 q_2} g_{q_1 q_2 \tilde{q}\tilde{q}'}(t)\left(i\Omega_{q_1 q_2 qq'} + \Gamma_{q_1 q_2 qq'}\right) = -\delta(t)\delta_{q\tilde{q}}\delta_{q'\tilde{q}'} \tag{4.2.8}$$

in terms of which one can easily express the linear response $\delta\rho_{qq'}(t)$:

$$\delta\rho_{qq'}(t) = -\frac{i}{\hbar}\sum_{\tilde{q}\tilde{q}'}\int_{-\infty}^{\infty} dt' g_{qq'\tilde{q}\tilde{q}'}(t-t')B_{\tilde{q}\tilde{q}'}\left(\rho_{\tilde{q}'} - \rho_{\tilde{q}}\right)F(t'). \tag{4.2.9}$$

Substituting (4.2.9) into (4.2.3) we are led to the usual way of writing down the time-dependent average value of a physical quantity $A(t)$ as the linear response to the perturbation (4.2.5): [144]

$$A(t) = \mathrm{Sp}\left(\hat{\rho}_0 \hat{A}\right) - \frac{1}{\hbar}\int_{-\infty}^{\infty} dt' G(t-t')F(t') \tag{4.2.10}$$

in which the required GF (4.2.1) is determined by the following expression:

$$G(t) = i\sum_{qq'\tilde{q}\tilde{q}'} A_{q'} g_{qq'\tilde{q}\tilde{q}'}(t) B_{\tilde{q}\tilde{q}'}\left(\rho_{\tilde{q}'} - \rho_{\tilde{q}}\right). \tag{4.2.11}$$

Together with Eqs. (4.2.7) and (4.2.8), Eq. (4.2.11) solves the given problem of finding the GF of a subsystem in the Markov approximation.

4.2.1. The exact solution for the exchange dephasing model

To begin with, we consider how Eqs. (4.2.7), (4.2.8), and (4.2.11) reproduce the main results provided by the known exchange dephasing model[147,148] as regards the spectral line shape for a high-frequency local vibration.[149,150]

The Hamiltonian of this model with a high-frequency mode ω_h and a single low-frequency exchange mode ω_l is given by Eq. (4.2.2) in which components have the following form:

$$\hat{H}^{(\mathrm{mol})} = \hbar\omega_h\left(b_h^+ b_h + 1/2\right) + \hbar\omega_l\left(b_l^+ b_l + 1/2\right) + \hbar\gamma b_h^+ b_h b_l^+ b_l\,,$$
$$\hat{H}_S = \sum_k \hbar\omega_k\left(b_k^+ b_k + 1/2\right), \qquad (4.2.12)$$
$$\hat{V} = \sum_k \hbar\left(\chi_k b_l^+ b_k + \chi_k^* b_l b_k^+\right).$$

The high-frequency mode is coupled through the anharmonicity coefficient γ with the low-frequency mode which is a resonant one due to harmonic interaction with the surface reservoir. For simplicity, the last term is written in the Gaitler-London approximation (compare with more general Eq. (4.1.8) where this restriction is absent). The required GF of the high-frequency mode can be obtained from the general Eq. (4.2.11) for $\hat{A} = b_h$, $\hat{B} = b_h^+$ and the GF $g_{qq'\tilde{q}\tilde{q}'}$ with $q = \{n_h, n_l\}$ ($b_h^+ b_h|n_h\rangle = n_h|n_h\rangle$, $b_l^+ b_l|n_l\rangle = n_l|n_l\rangle$):

$$G(t) = i\sum_{n_l n_l'} g_{n_l n_l'}(t)\rho_{n_l'}\,, \quad g_{n_l n_l'}(t) = g_{1n_l,0n_l,1n_l',0n_l'}(t)\,. \qquad (4.2.13)$$

In this formula, we have used the inequality $\hbar\omega_h >> T$ which makes it sufficient to consider only the states with $n_h = 0$ and 1, and also to put $\rho_{1n_l} = 0$ (the quantity ρ_{0n_l} is denoted by ρ_{n_l}). Calculating the matrix elements (4.2.7) by Eqs. (4.2.12) and substituting them in Eq. (4.2.8) we arrive at:

$$\begin{aligned}
&\frac{\partial}{\partial t} g_{n_l n_l'}(t) + \{i(\omega_h + \gamma n_l) + [n_l + (n_l+1)\xi]w_0\} g_{n_l n_l'}(t) \\
&- (n_l+1)w_0 g_{n_l+1,n_l'}(t) - n_l\xi w_0 g_{n_l-1,n_l'}(t) = -\delta(t)\delta_{n_l n_l'} \\
&w_0 = \eta(\omega_l)/(1-\xi), \quad \eta(\omega_l) = 2\pi\sum_k |\chi_k|^2\delta(\omega_l - \omega_k) \\
&\xi = \exp(-\hbar\omega_l/T).
\end{aligned} \qquad (4.2.14)$$

The structure of the left-hand side of Eq. (4.2.14) is the same as the homogeneous equation of Ref. 148 for the matrix elements $\rho_{0n_l,1n_l}$ of the reduced density matrix of the subsystem considered. A numerical solution of that equation was given in Ref. 151. Exact analytical expressions for the spectral line shape of a high-frequency local oscillation were obtained in terms of this model by the generating-function[149] and temperature-GF[150,152] methods. (The latter approach leads to an integral equation representing the boson version of the equation written down many years ago by Nozières and De Dominicis[153] to solve the x-ray edge problem). Using the approach of Ref. 149 one can obtain an analytical expression for the GF (4.2.13) in the case of several low-frequency exchange modes ω_j (j = 1,2,...,p) which interact with one another neither directly nor through the reservoir:

$$G(t) = -i\theta(t)\exp(-i\omega_l t)\prod_{j=1}^{p}\frac{\left(1-\tilde{\xi}_j\right)\exp\left(-i\lambda_j t\right)}{1-\tilde{\xi}_j\exp\left(-i\nu_j t\right)}. \tag{4.2.15}$$

Here we have

$$\tilde{\xi} = \left(\frac{1-\mu}{1-\xi\mu}\right)^2\xi,\quad \lambda = -\frac{i\xi\nu}{1-\xi}(1-\mu),\quad \nu = -\frac{i\eta\left(1-\xi\mu^2\right)}{(1-\xi)\mu},$$
$$\mu = \frac{1-\xi}{2\xi}\left[\frac{1+\xi}{1-\xi}+i\frac{\gamma}{\eta}-\left(1+2i\frac{1+\xi}{1-\xi}\frac{\gamma}{\eta}-\frac{\gamma^2}{\eta^2}\right)^{1/2}\right] \tag{4.2.16}$$

(to simplify the notation, we have omitted the index j of the quantities ξ, γ, η, μ, $\tilde{\xi}$, λ, and ν).

Eq. (4.2.15) can be simplified in the practically important case of p degenerate exchange modes which is realized, e.g., for molecular complexes with hydrogen bonds. The high-frequency mode ω_h is represented by the valence or deformation vibration of a hydrogen atom which is anharmonically coupled with four (p=4) degenerate libration modes whose low frequencies are caused by the relatively weak interaction between two molecular fragments through the hydrogen bond (Fig. 4.2). On expanding the denominator in (4.2.15) in a series and integrating over the time, the spectral function of the high-frequency vibration near ω_h for p degenerate exchange modes takes the form

$$S(\omega) = -\frac{1}{\pi}\operatorname{Im}G(\omega),$$
$$G(\omega) = \frac{\left(1-\tilde{\xi}\right)^p}{(p-1)!}\sum_{n=0}^{\infty}\frac{(n+p-1)!}{n!}\frac{\tilde{\xi}^n}{\omega-\omega_h-p\lambda-n\nu}. \tag{4.2.17}$$

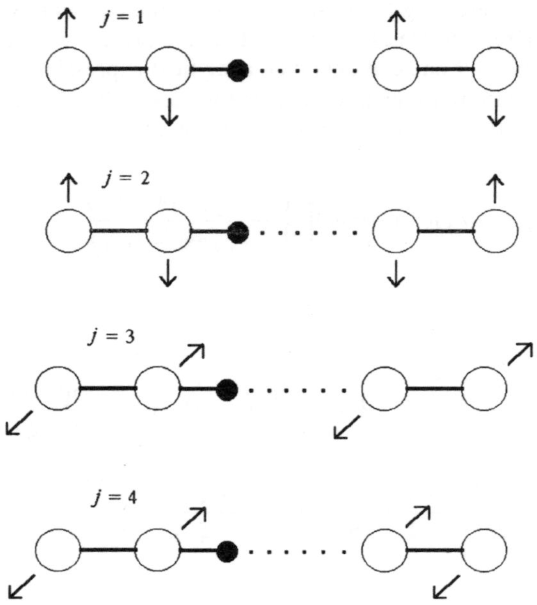

Fig. 4.2. Low-frequency modes of librational vibrations of a molecular complex with a hydrogen bond.

For $p = 1$, Eq. (4.2.17) reduces to the results obtained in Refs. 149 and 150 and in the high-temperature limit ($T >> \hbar\omega_j$} gives, at $\eta << \gamma$, a simple formula[121,154] for the strong one-sided temperature broadening of the spectral lines of complexes with hydrogen bonds (see Fig. 4.3):

$$S(\omega) = \frac{\hbar\omega_l}{(p-1)!|\gamma|T}\theta(z)z^{p-1}e^{-z}, \quad z = \frac{\hbar\omega_l}{\gamma T}(\omega - \omega_h). \tag{4.2.18}$$

Note that the molecules in gaseous phase do not exchange energy with the thermostat so that expression (4.2.17) corresponding to the limiting case $\eta_j \to 0$ is consistent with the concept of hot electrons (with the frequencies $\omega_h + \sum\gamma_j\xi_j/(1-\xi_j)$) which accounts for the fine structure of spectral lines of gaseous H-bond complexes.[155] The one-sided broadening of the spectral line 2Γ is proportional to the shift of its maximum $\Delta\omega$ for various H-bond complexes in different phases ($2\Gamma \approx \kappa\Delta\omega$, $\kappa \approx 1 \div 1.5$), both quantities increasing with the temperature quadratically at small T and linearly at large T, in accordance with the approximated temperature dependence $\omega_h + \sum\gamma_j\xi_j/(1-\xi_j)$ on different temperature intervals. For instance, *tert*-butanol manifests the characteristic dependence $2\Gamma \approx 1.32\times10^{-3}\ T^2$ (Γ is measured in cm^{-1}) at $15 < T < 290$ K ($\kappa \approx 1.0$ at $T < 150$ K and $\kappa \approx 1.4$ at $T > 150$ K).[156,157]

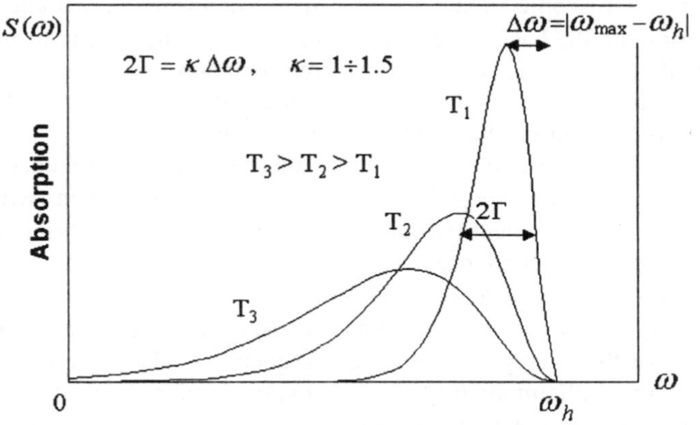

Fig. 4.3. One-sided temperature broadening of spectral lines typical of H-bond complexes.

Formula (4.2.18) represents an envelope of spectral function (4.2.17) at high temperatures ($T > \hbar\omega_l$) when integration instead of summation over all n_l is possible in Eq. (4.2.13). The typical parameter values in Eq. (4.2.18) for complexes with hydrogen bonds are $\omega_l \sim 30\ cm^{-1}$, $\gamma \sim 3\ cm^{-1}$, and $p = 4$.[155-157] The shift of the spectral function maximum specified by Eq. (4.2.18), $\Delta\omega = |\omega_{max} - \omega_h| = (p-1)\gamma T/\hbar\omega_l$, is proportional to the temperature; at $T = 300$ K, the value $\Delta\omega \sim 60\ cm^{-1}$ agrees with the experimental evidence. The proportionality factor κ is defined as a difference of two roots of the equation $x = 2^{(1-p)^{-1}} e^{x-1}$: $\kappa = x_2 - x_1$ ($x_1 < x_2$), and varies from 2.45 at $p = 2$ to 0.788 at $p = 10$, the calculated value $\kappa = 1.38$ at $p = 4$ for *tert*-butanol closely coinciding with the measured one. At the temperature $T = 290$ K corresponding to the phase transition of *tert*-butanol from the liquid to the crystalline phase, the spectral line narrows jump-like by about 30 cm^{-1} [156,157] on account of the drastic enhancement of η exchange in crystalline phase, this effect being accompanied with the accordingly decreased shift $\Delta\omega$ of the spectral function maximum. On adsorption of H-bond complexes, the fourfold degeneracy of low-frequency libration modes should be removed, as the substrate influences atomic displacements in adsorbed molecules unequally in two perpendicular directions. As a result, four pairwise-degenerate libration modes originate and the coefficient κ grows from 1.38 to 2.45 together with the degeneration removal parameter.[158]

The exact Eq. (4.2.17) takes into account the effect of the reservoir (the condensed phase) on the spectral line shape through the parameter η. Consideration of a concrete microscopic model of the valence-deformation vibrations makes it possible to estimate the basic parameters γ and η of the theory and to introduce the exchange mode anharmonicity caused by a reorientation barrier of the deformation vibrations; thereby, one can fully take advantage of the GF representation in the form (4.2.11) which allows summation over a finite number of states.

Indeed, torsional vibrations are strongly anharmonic; they are characterized by well-defined values of the energy barriers ΔU separating equivalent (with a rotation angle $\varphi = 2\pi$) or nonequivalent (with $\varphi = 2\pi/p$, $p = 2, 3, ...$) equilibrium orientations of the molecules in the condensed phase. One can talk about torsional vibrations only for sufficiently low values of the energy $\varepsilon < \Delta U$, whereas for $\varepsilon \geq \Delta U$ stochastic reorientation processes dominate causing a spectral line broadening by the magnitude of the average reorientation frequency.[159] The simplest way to take into account the anharmonicity of the torsional vibrations thus consists in considering a limited number of orientational states (with $\varepsilon < \Delta U$), which leads to the observed Arrhenius-type temperature dependence of the line width dictated by the factor $\exp(-\Delta U/T)$. The pre-exponential factor depends on the relations between the parameters of the system which were established in the classical consideration.[160]

The problems discussed here are closely related to the problem of calculating the rates at which a particle leaves a potential well and which govern the rates of chemical reactions. The most consistent description of low-temperature chemical reactions that included tunnelling and dissipation processes was given in Ref. 161. We shall be interested only in the thermally activated contribution which dominates for many systems at not too low temperatures.

4.2.2. Valence-deformation vibrations of a molecular subsystem in condensed phase

As a simple model which takes into account valence and deformation vibrations of a molecule imbedded in the condensed phase, we consider a diatomic molecule with two degrees of freedom corresponding to valence (in the radial variable r) and torsional (in the angular variable φ) vibrations:[146]

$$\hat{H}^{(\mathrm{mol})} \equiv \hat{H}_{r\varphi} = -\frac{\hbar^2}{2m^*}\frac{\partial^2}{\partial r^2} - \frac{\hbar^2}{2mr^2}\frac{\partial^2}{\partial \varphi^2} + U(r,\varphi),$$
$$U(r,\varphi) \approx \frac{1}{2}m^*\omega_h^2(r-r_0)^2 + U_{\mathrm{anh}}(r) + \frac{1}{2}\Delta U(r)(1-\cos p\varphi), \quad p = 1,\ 2,\ \ldots, \tag{4.2.19}$$

where m is the mass of the atom involved in angular deformations, m^* is the reduced mass of the molecule, r_0 is the equilibrium radial distance, and the approximate expression for the potential $U(r,\varphi)$ describes the valence vibration of the frequency ω_h and the hindered rotation of the molecule in the p-well potential of surrounding atoms. A reduced Hamiltonian was used in Refs. 61 and 121 for the description of the vibrational states of surface groups of atoms with rotational degrees of freedom and it can serve as a microscopic model which takes account of the basic factors governing the spectral line shape for the high-frequency valence vibration of a molecule coupled with the low-frequency deformation mode.

The radial variable r in the operator of angular kinetic energy and the radial dependence of the reorientation barrier ΔU(r) give rise to the coupling of the radial and the deformation motions. In the framework of the exchange dephasing model, we take into account the biquadratic anharmonic term $(r - r_0)^2\varphi^2$ arising from the operator of angular kinetic energy and describing the self-scattering of the local vibration coupled with the low-frequency modes[162,163] (contributions of other anharmonic terms will be discussed in the next section). Then we arrive at:

$$\hat{H}_{r\varphi} = \hbar\omega_h\left(b_h^+b_h + \frac{1}{2}\right) + \hat{H}_\varphi + \frac{3}{2}\frac{m}{m^*}\frac{\hbar\omega_l^2}{\omega_h}\varphi^2 b_h^+b_h,$$
$$\hat{H}_\varphi = -\frac{\hbar^2}{2mr_0^2}\frac{\partial^2}{\partial\varphi^2} + \frac{1}{2}\Delta U(r_0)(1-\cos p\varphi), \tag{4.2.20}$$

where $\omega_l \approx p\left(\Delta U/2mr_0^2\right)^{1/2}$ is the characteristic frequency of torsional vibrations.

In Fig. 4.4 we schematically show the eigenvalues $\varepsilon_{n_h\sigma}$ of the Hamiltonian $\hat{H}_{r\varphi}$ which depend on the radial, n_h, and the deformation, σ, quantum numbers. For $\varepsilon_{0\sigma} << \Delta U$ the levels are grouped so that the gaps between groups are approximately the same and equal to $\hbar\omega_l$ whereas the tunneling splitting which occurs for $p > 1$ in each group of p levels is exponentially small in the parameter $4\Delta U/\hbar\omega_l$.[61,121]

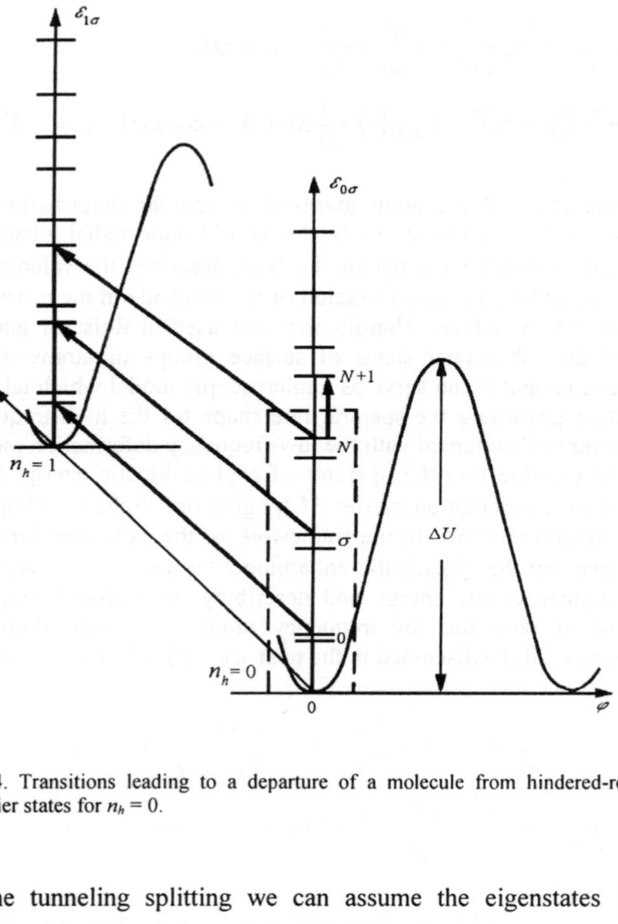

Fig. 4.4. Transitions leading to a departure of a molecule from hindered-rotation-subbarrier states for $n_h = 0$.

Neglecting the tunneling splitting we can assume the eigenstates $|0\sigma\rangle$ to be localized in the wells of the deformation potential so that

$$\begin{aligned} &\varepsilon_{0\sigma} = \hbar\omega_l(\sigma + 1/2), \quad \sigma = 0,\ 1,\ \ldots << N + 1 \approx \Delta U(r_0)/\hbar\omega_l\,, \\ &\langle 0\sigma'|\cos\varphi|0\sigma\rangle = \delta_{\sigma'\sigma}\,, \\ &\langle 0\sigma'|\sin\varphi|0\sigma\rangle = \left(\hbar/2mr_0^2\omega_l\right)^{1/2}\left[\sigma^{1/2}\delta_{\sigma'\sigma-1} + (\sigma+1)^{1/2}\delta_{\sigma'\sigma+1}\right]. \end{aligned} \tag{4.2.21}$$

(Tunnel relaxation of orientational states in the phonon field of a substrate is considered in Appendix 2). When a molecule has a single equilibrium orientation (p = 1) the deformation potential is also characterized by a well-defined barrier ΔU which separates the equivalent minima. That is why, the subsystem Hamiltonian (4.2.12) used in the exchange dephasing model[147,148] with

$$\gamma = \frac{3\hbar\omega_l}{2mr_0^2\omega_h}, \quad \varphi = \left(\frac{\hbar}{2mr_0^2\omega_l}\right)^{1/2}\left(b_l + b_l^+\right) \tag{4.2.22}$$

correctly describes only the subbarrier deformation states (the operators b_l and b_l^+ are now defined by the standard expressions for the matrix elements of the subbarrier states only). When we describe the high-frequency response with regard to the molecular orientation, we must put the operators $\hat{A}$ and $\hat{B}$ in the GF (4.2.1) as:

$$\hat{A} = b_h \cos\varphi, \quad \hat{B} = b_h^+ \cos\varphi\,. \tag{4.2.23}$$

By virtue of the properties of Eq. (4.2.21), the matrix elements $A_{q'q}$ and $B_{\tilde{q}'\tilde{q}}$ in Eq. (4.2.11) will be diagonal in the quantum numbers σ only for the subbarrier states, and the operator $\hat{V}$ takes the form (4.2.12) for those states.

Indeed, it is shown in Ref. 164 (see also Ref. 61) that the interaction of a reorienting molecule with a solid matrix (in which it is embedded due to the rigid coupling through only one of its atoms) can be written as the energy of the d'Alembert force, $-m\ddot{\mathbf{u}}$, in the noninertial frame of reference connected to the molecular center of mass which undergoes the acceleration $\ddot{\mathbf{u}}$ due to vibrations of the solid: $V = m\mathbf{r}\cdot\ddot{\mathbf{u}}$. Expressing the deformation vector $\mathbf{u}$ in terms of the second quantization operators of phonons of the reservoir and using the fact that the operators b_l and b_l^+ arise due to the presence of $\sin\varphi$ in the vector $\mathbf{r} = r_0(\cos\varphi, \sin\varphi)$, the operator $\hat{V}$ for the subbarrier states can be written as in Eq. (4.2.12) with

$$\chi_k = \frac{1}{2}\left(\frac{m\omega_k^3}{\rho V\omega_l}\right)^{1/2}, \tag{4.2.24}$$

where ρ is the density of the medium and V is the volume of the main area. In the case of a Debye spectrum for phonons of the reservoir which are characterized by the average sound velocity c or the Debye frequency ω_D and the mass M of the unit cell, the width of the resonant deformation mode takes the form[164]

$$\eta = \frac{m\omega_l^4}{4\pi\rho c^3} = \frac{3\pi}{2}\frac{m}{M}\left(\frac{\omega_l}{\omega_D}\right)^3 \omega_l \,. \tag{4.2.25}$$

(For an alternative derivation of this formula, see Appendix 1). For real systems, we have $m < M$, $(\omega_l/\omega_D)^3 << 1$, $\hbar/2mr_0^2\omega_l << 1$ and the estimates (4.2.22) and (4.2.25) satisfy the conditions $\eta, \gamma << \omega_l$ formulated for the Markov approximation.

Transitions between superbarrier states ($\sigma > N$, see Fig. 4.4) involve nondiagonal in σ elements of the density matrix which give, at $\Delta U > T$, a small contribution to the spectral function for frequencies of the order $\omega_h + (\varepsilon_{0\sigma} - \varepsilon_{0\sigma'})/\hbar$ $(\sigma \neq \sigma')$. As we are interested in the spectral function at frequencies near ω_h, we can neglect this contribution.

Thus, the required expression for the GF of the high-frequency mode taking into account the anharmonic coupling of the latter with the exchange deformation mode (characterized by a well-defined value of the reorientation barrier ΔU) takes the form (4.2.13) with restricted summation over the quantum numbers $n_l = \sigma = 0, 1, \ldots << N$ of the subbarrier states. It is then expedient to rewrite Eq. (4.2.14) in the following form:

$$\frac{\partial}{\partial t} g_{\sigma\sigma'}(t) + i(\omega_h + v_\sigma) g_{\sigma\sigma'}(t) + \sum_{\sigma''=0}^{N} W_{\sigma\sigma''} g_{\sigma''\sigma'}(t) = -\delta(t)\delta_{\sigma\sigma'} \,, \tag{4.2.26}$$

$$v_\sigma = i\gamma\sigma + w_0(N+1)\xi\delta_{\sigma N} \,, \quad \xi = \exp(-\hbar\omega_l/T) \,, \tag{4.2.27}$$

$$\begin{aligned} W_{\sigma\sigma'} = w_0\{ & [\sigma + (\sigma+1)\xi - (N+1)\xi\delta_{\sigma N}]\delta_{\sigma'\sigma} \\ & - (\sigma+1)\delta_{\sigma',\sigma+1} - \sigma\xi\delta_{\sigma',\sigma-1}\}, \quad \sigma, \sigma' = 0, 1, \ldots, N \,, \end{aligned} \tag{4.2.28}$$

which corresponds to the Pauli equation with the transition rates perturbed relative to $W_{\sigma\sigma'}$. The perturbations v_σ determine the rate at which the molecule leaves the subbarrier states (see the area enclosed by the dashed lines in Fig. 4.4) and consist of two contributions: departures due to the anharmonic coupling of the valence and the torsional vibrations and those due to the strong anharmonicity of the torsional vibrations (the anharmonicity is determined by the magnitude of the barrier ΔU in the present model). The latter contribution proportional to $w_0(N+1)\xi$ was neglected in Ref. 151, since an equation like Eq. (4.2.26) with finite N was used only for a numerical approximation of the case $N \to \infty$. The next section will cover elucidation of properties for the GF of the Pauli equation with unperturbed transition rates $W_{\sigma\sigma'}$ which satisfy the principle of detailed balance; on this basis, a perturbation theory in small corrections to the transition rates will be constructed, and the results will be considered in detail for a model with ν_σ and $W_{\sigma\sigma'}$ of the form (4.2.27) and (4.2.28).

4.2.3. Perturbation theory for the transition rates in the Pauli equation

We consider a subsystem characterized by the states with the energies ε_σ and the rates $w_{qq'}$ for transitions from a state q' into a state q that satisfy the principle of detailed balance.

$$w_{qq'} = \frac{\rho_q}{\rho_{q'}} w_{q'q}, \quad \rho_q = \frac{\exp(-\varepsilon_q/T)}{\sum_q \exp(-\varepsilon_q/T)}. \tag{4.2.29}$$

The Pauli equation[165] for the probability $p_q(t)$ of finding the subsystem in the state q at time t is conveniently written to perform further transformations:

$$\frac{d}{dt} p_q(t) + \sum_{q'} W_{qq'} p_{q'}(t) = 0, \quad W_{qq'} = \delta_{qq'} \sum_{q''} w_{q''q} - w_{qq'}. \tag{4.2.30}$$

The principle of detailed balance which is also valid for the quantities $W_{qq'}$ enables the diagonalization of the nonsymmetric matrix $W_{qq'}$ with nonnegative elements:

$$\sum_{q'} W_{qq'} C_{q'\nu} = \mu_\nu C_{q\nu}, \tag{4.2.31}$$

where the eigenvalues μ_ν are also nonnegative and the orthonormalization relations for the matrix elements $C_{q\nu}$ are defined with weights ρ_q:

$$\sum_\nu C_{q\nu} C_{q'\nu} = \rho_q \delta_{qq'}, \quad \sum_q \rho_q^{-1} C_{q\nu} C_{q'\nu'} = \delta_{\nu\nu'}. \tag{4.2.32}$$

If we denote the initial probabilities for state occupations at time t=0 by $p_q(0)$, the solution of Eq. (4.2.30) takes the following form:

$$p_q(t) = \sum_{q'} \rho_{q'}^{-1} p_{q'}(0) \sum_{\nu} C_{q\nu} C_{q'\nu} \exp(-\mu_\nu t). \tag{4.2.33}$$

The quantities introduced have a number of properties necessary for thermodynamic equilibrium to establish between the subsystem and the reservoir at $t \to \infty$. First of all, we note that by virtue of the definition (4.2.30) the summation of the matrix elements $W_{qq'}$ over the first index gives zero. Therefore, summation over q of both sides of Eq. (4.2.31) makes the product $\mu_\nu \sum_q C_{q\nu}$ vanish. From the linear independence of the rows C_q of the transformation it follows that there exists at least one eigenvalue μ_ν equal to zero. We denote the corresponding index ν by zero, and then $\mu_0 = 0$ and $\sum_q C_{q\nu} = 0$ for $\nu \neq 0$. Summing the first of Eqs. (4.2.32) over q and using the fact that $\sum_q \rho_q = 1$ we find

$$\mu_0 = 0, \quad C_{q0} = \rho_q, \quad \sum_q C_{q\nu} = \delta_{\nu 0}. \tag{4.2.34}$$

These properties lead to the physically obvious consequences of the solution of Eq. (4.2.33):

$$\sum_q p_q(t) = \sum_q p_q(0) = 1,$$

$p_q(t) = \rho_q$ for $p_q(0) = \rho_q$, and $p_q(\infty) = \rho_q$ for arbitrary initial conditions.

The frequency Fourier component $g_{qq'}^{(0)}(\omega)$ of the GF of the unperturbed Pauli equation (4.2.30) (satisfying Eq. (4.2.30) with $-\delta(t)\delta_{qq'}$ on the right-hand side) has, in view of Eq. (4.2.32), a pole at $\omega = 0$:

$$g_{qq'}^{(0)}(\omega) = -i \sum_{\nu} \frac{C_{q\nu} C_{q'\nu} \rho_{q'}^{-1}}{\omega + i\mu_\nu} = -i \frac{\rho_q}{\omega} - \varepsilon_{qq'}(\omega). \tag{4.2.35}$$

Here $\varepsilon_{qq'}(\omega)$ is determined by the general expression for $g^{(0)}_{qq'}(\omega)$ in which the term with $\mu_\nu = 0$ is excluded from the sum over ν, so that, for instance, $\varepsilon_{qq'}(0)$, with the weight factor ρ taken into account, is a pseudoinverse matrix with respect to $W_{qq'}$ and satisfies the following identities:

$$\sum_{q''} W_{qq''}\varepsilon_{q''q'}(0) = \delta_{qq'} - \rho_q, \quad \sum_q \varepsilon_{qq'}(0) = 0. \tag{4.2.36}$$

We now consider perturbations of the transition rates, adding the diagonal in q contribution $\nu_q\delta_{qq'}$ to $W_{qq'}$. The perturbed GF $g_{qq'}(\omega)$ will then be related with the unperturbed one through the Dyson equation[146]

$$g_{qq'}(\omega) = g^{(0)}_{qq'}(\omega) + \sum_{q''} g^{(0)}_{qq''}(\omega)\nu_{q''}g_{q''q'}(\omega). \tag{4.2.37}$$

In the second-order perturbation theory in ν_q, the required retarded GF (25) is equal to

$$G(t) = -i\theta(t)\exp(-\tilde{\Gamma}t), \quad \tilde{\Gamma} = \sum_q \nu_q\rho_q - \sum_{qq'} \nu_q\varepsilon_{qq'}(0)\nu_{q'}\rho_{q'} \tag{4.2.38}$$

and determines, accurate to the factor $-i\theta(t)$, the probability that the subsystem leaves the given group of states (e.g., those enclosed by the dashed line in Fig. 4.4). The quantity $\tilde{\Gamma}$ acquires the meaning of a generalized leaving rate, since ν_q and $\tilde{\Gamma}$ can take on complex values. The spectral function corresponding to Eqs. (4.2.13) and (4.2.26) becomes Lorentzian:

$$S(\omega) = -\frac{1}{\pi}\operatorname{Im}\frac{1}{\omega - \omega_h + i\tilde{\Gamma}}. \tag{4.2.39}$$

The first term in expression (4.2.38) for $\tilde{\Gamma}$ has a simple physical meaning: it sums the perturbed leaving rates from each level of the subsystem taking into account the equilibrium probabilities for their occupation. On the other hand, the second term depends on the unperturbed rates for transitions between states of the subsystem and is inversely proportional to them by virtue of the definition of $\varepsilon_{qq'}(0)$.

We evaluate the pseudoinverse matrix $\varepsilon_{qq'}(0)$ of the transition rates (4.2.28) between $N+1$ low-energy states of a harmonic oscillator which interacts resonantly with a phonon reservoir. With this aim in view, it is necessary to: 1) write $g^{(0)}_{qq''}(\omega)$ as the ratio of the appropriate cofactor to the determinant of the matrix $i\omega\delta_{qq'} - W_{qq'}$; 2) expand this quantity in a Laurent series in ω; then the principal part of the Laurent series gives the first term of the right-hand side of Eq. (4.2.35) and the terms of zeroth degree in ω from the regular part of the Laurent series determine the required $\varepsilon_{qq'}(0)$ matrix; 3) calculate resulting determinants and their derivatives with respect to ω for $\omega = 0$ using relatively simple recurrence relations obtained by expanding determinants of a quasi-triangular shape (like the matrix (4.2.28) for $W_{qq'}$) with respect to an arbitrary row or column. As a result, we get

$$\varepsilon_{qq'}(0) = \frac{\rho_q}{\rho_{q'}}\varepsilon_{q'q}(0) = \frac{\rho_q}{w_0(1-\xi)}\left[\sum_{k=1}^{q'}\frac{\xi^{-k}-1}{k} + \sum_{k=q+1}^{N}\frac{1-\xi^{N+1-k}}{k} - \frac{1}{1-\xi^{N+1}}\sum_{k=1}^{N}\frac{\left(1-\xi^k\right)\left(1-\xi^{N+1-k}\right)}{k}\right], \quad q \ge q', \quad \rho_q = \frac{1-\xi}{1-\xi^{N+1-k}}\xi^q. \tag{4.2.40}$$

Here q, $q' = 0, 1, \ldots, N$ and it is implied for $q = q' = 0$ and $q = N$ that sums vanish if the upper limit of summation smaller than the lower one. It is easy to check that the expression (4.2.40) derived here obeys the identities (4.2.36).

As far as our concern is with the second-order perturbation theory specified by Eqs. (4.2.38) and (4.2.39) and valid for $\gamma \ll \eta$ and $\xi^{N+1} \ll 1$, we can now find the half-width Γ of the spectral function and the shift $\Delta\omega$ of the maximum relative to ω_h for the problem formulated by formulae (4.2.26)-(4.2.28):[146]

$$\Gamma = \mathrm{Re}\,\widetilde{\Gamma} = (N+1)\eta\frac{\xi^{N+1}}{1-\xi^{N+1}}A_N(\xi) + \frac{\gamma^2}{\eta}\frac{\xi}{(1-\xi)^2}B_N(\xi), \tag{4.2.41}$$

$$\Delta\omega = \mathrm{Im}\,\widetilde{\Gamma} = \gamma\frac{\xi}{1-\xi}C_N(\xi), \tag{4.2.42}$$

where

$$\begin{aligned}
A_N(\xi) &= 1 - \frac{(N+1)\xi^{N+1}}{\left(1-\xi^{N+1}\right)^2} f_N(\xi), \quad f_N(\xi) = \sum_{k=1}^{N} \frac{\left(1-\xi^k\right)^2}{k\xi^k}, \\
B_N(\xi) &= 1 - \frac{(N+1)(1-\xi)^2 \xi^N}{1-\xi^{N+1}} \left[A_N(\xi) + \frac{\xi}{1-\xi} + D_N(\xi) \right], \\
C_N(\xi) &= 1 - \frac{(N+1)(1-\xi)\xi^N}{1-\xi^{N+1}} \left[2A_N(\xi) - 1 + D_N(\xi) \right], \\
D_N(\xi) &= \frac{2}{1-\xi^{N+1}} \left[N - \frac{\xi\left(1-\xi^N\right)}{1-\xi} \right].
\end{aligned} \tag{4.2.43}$$

For a single subbarrier level we have $A_0(\xi) = 1$, $B_0(\xi) = C_0(\xi) = 0$, and the half-width of the spectral function is determined solely by the reorientation rate, which is equal to the transition rate to the first excited state of deformation vibrations:[164]

$$\Gamma = \eta\, n(\omega_l), \quad n(\omega_l) = \left[\exp(\hbar\omega_l/T) - 1\right]^{-1}. \tag{4.2.44}$$

For two subbarrier levels ($N = 1$), Eqs. (4.2.43) represent, at $\xi << 1$, an approximated result of the two-level problem that holds true for any values of ξ:

$$\begin{aligned}
A_1(\xi) &= \frac{1}{4\xi^2}(1+\xi)\left[1 + 3\xi - \left(1 + 6\xi + \xi^2\right)^{1/2}\right], \\
B_1(\xi) &= \frac{(1-\xi)^2}{\left(1+6\xi+\xi^2\right)^{3/2}}, \quad C_1(\xi) = \frac{1-\xi}{2\xi}\left[1 - \frac{1+\xi}{\left(1+6\xi+\xi^2\right)^{1/2}}\right].
\end{aligned} \tag{4.2.45}$$

Finally, in the other limiting case, $N \to \infty$, the functions $B_N(\xi)$ and $C_N(\xi)$ are approximately equal to unity, and Eqs. (4.2.41) and (4.2.42) reduce to the results of Ref. 151 for $\gamma << \eta$.

We draw attention to the fact that in the first-order perturbation theory we have $A_N(\xi) = 1$ and $B_N(\xi) = 0$ in Eq. (4.2.41) for any N, and the expression for the leaving rate of the molecule from the subbarrier states reduces to two well-known special cases. The first of these corresponds to the low-temperature limit $\xi \to 0$ for which[166]

$$\eta = w_0(1-\xi) \approx w_0, \quad \Gamma = (N+1) w_0 \exp(-\Delta U/T).$$

The second special case corresponds to the classical limit $1-\xi \to \hbar\omega_l/T \to 0$ and gives the Kramers' result:[160]

$$\Gamma = w_0(\Delta U/T)\exp(-\Delta U/T),$$

in which the parameter w_0 serves as a "viscosity" coefficient. In the second-order perturbation theory, the coefficient $A_N(\xi)$ refines the first-order result. For $N >> 1$ and $\xi \le 0.5$, we derive the asymptotic expression from Eq.(4.2.43):

$$A_N(\xi) \approx 1-\xi/(1-\xi).$$

The main results of the present section which are helpful in describing spectra of high-frequency local vibrations of molecular subsystems in the condensed phase are contained in Eqs. (4.2.17), (4.2.18), and (4.2.41) to (4.2.43) which involve two parameters, γ and η. According to the estimates of Ref. 155, molecular complexes with hydrogen bonds are characterized by the anharmonic coupling coefficient $\gamma \sim 3$ cm^{-1} and the frequencies of the libration vibrations $\omega_l \sim 30$ cm^{-1} much less than the Debye frequency ω_D,. This fact leads to the inequality $\eta << \gamma$ thus corroborating the validity of Eq. (4.2.18), in agreement with the experimental data of Refs. 156 and 157 (a discussion of this problem is given in Refs. 121 and 154).

Estimation of the parameters γ and η for surface groups of atoms using Eqs. (4.2.22) and (4.2.25) or similar relations in Ref. 151 shows that $\gamma/\eta \sim 0.1$. If we put $\gamma/\eta = 0.1$ in (4.2.41), the reorientation contribution to the broadening becomes dominating over the anharmonic contribution with the coefficient γ, starting at temperatures $T > 0.5\ \hbar\omega_l$ for $N = 3$ or $T > 0.9\ \hbar\omega_l$ for $N = 5$. Assuming that the values of the barriers ΔU for CO bridge groups on Ni(111) correspond to $N \le 3$ and involving the realistic estimates $\omega_l \approx 184$ cm^{-1}, $\omega_h \approx 1900$ cm^{-1}, $\gamma \sim 1$ cm^{-1}, $\eta \sim 30$ cm^{-1}, the reorientation contribution to Eq. (4.2.41) can explain the previously observed[151] spectral line broadening for the valence CO vibrations. Apart from the contribution (4.2.42) from the fourth-degree anharmonicity $(r-r_0)^2\varphi^2$, even the third-degree anharmonicity of the form $(r-r_0)\varphi^2$ and also other low-frequency vibrational modes of the substrate[167] will give a comparable contribution to the shift of the maximum of the same line. Neglecting the other low-frequency modes and taking into account only a single harmonic ($N \to \infty$) mode, one would overestimate, by an order of magnitude, the value of the only anharmonic parameter γ in the description of the observed spectra.[150,151]

The purely reorientational broadening mechanism with a single threefold quasi-degenerate subbarrier level is characteristic of the valence vibration spectral line for OH groups on SiO_2 surface. Equation (4.2.22) describes the observed temperature dependence of the line halfwidth for $\omega_l \approx 200$ cm^{-1} and $\eta \approx 4$ cm^{-1} (for this system, $\gamma \approx 4$ cm^{-1} holds and in Eq. (4.2.22) we have $\gamma^2/\eta \approx 0.2$ cm^{-1}, which is much less than the value η).[61,121,164] Eqs. (4.2.41), (4.2.42), and (4.2.45) are applicable to the system of OD groups on SiO_2 surface, since in this case one can put the number of subbarrier levels equal to two due to the doubled mass of the reorienting atom.

Importantly, the value of the results gained in the present section is not limited to the application to actual systems. Eq. (4.2.11) for the GF in the Markov approximation and the development of the perturbation theory for the Pauli equation which describes many physical systems satisfactorily have a rather general character. An effective use of the approaches proposed could be exemplified by tackling the problem on the rates of transitions of a particle between locally bound subsystems. The description of the spectrum of the latter considered in Ref. 135 by means of quantum-mechanical GF can easily be reformulated in terms of the GF of the Pauli equation.

4.3. Generalization of the exchange dephasing model to various cases of anharmonic coupling between high-frequency and low-frequency modes, and to the case of collectivized excitations in adsorbate

In Sec. 4.2 the exchange dephasing model was considered which takes into account only the biquadratic anharmonic coupling between the local and low-frequency vibrations (see Eq.(4.2.12). The need for due regard to cubic terms of anharmonic coupling was pointed out more than once.[1,168-169] First, the contribution of the cubic anharmonicity to relaxation processes is of particular significance in the cases when the doubled frequency of a resonance molecular mode is found close to the frequency of the local vibration, as for instance, in the system D/C(111).[170,171] Second, the cubic anharmonicity also contributes to the dephasing mechanism of line broadening and results in the renormalized constant of the biquadratic anharmonicity.[1] Such a renormalization of the quartic anharmonicity coefficient in terms of the cubic one was invoked by Ivanov, Krivoglaz *et al.* for consideration of high-frequency vibrations in crystals as early as in the sixties.[172,173] The necessity of taking into account cubic corrections in the processes involving four vibrational excitations is easy to comprehend from the analogy with an anharmonic oscillator treated in the perturbation theory. It is common knowledge that contributions from the quartic anharmonicity in the first-order perturbation theory and from the cubic anharmonicity in the second-order perturbation theory are of the same order of magnitude. Vibrational dephasing and relaxation in four-phonon processes are determined by the squared quartic anharmonicity coefficient. Therefore, the fourth-order perturbation theory is needed to adequately describe these processes.

Intermolecular lateral interactions and resulting collectivized vibrations of individual adsorbed molecules greatly add to the complexity of description for local vibrational excitations in adsorbates. Fig. 4.5 schematically demonstrates that these interactions on a simple planar lattice of adsorbed molecules which vibrate with high (ω_h) and low (ω_l) frequencies lead to the emergence of the corresponding energy bands, with energy levels classified by the wave vector **K**.

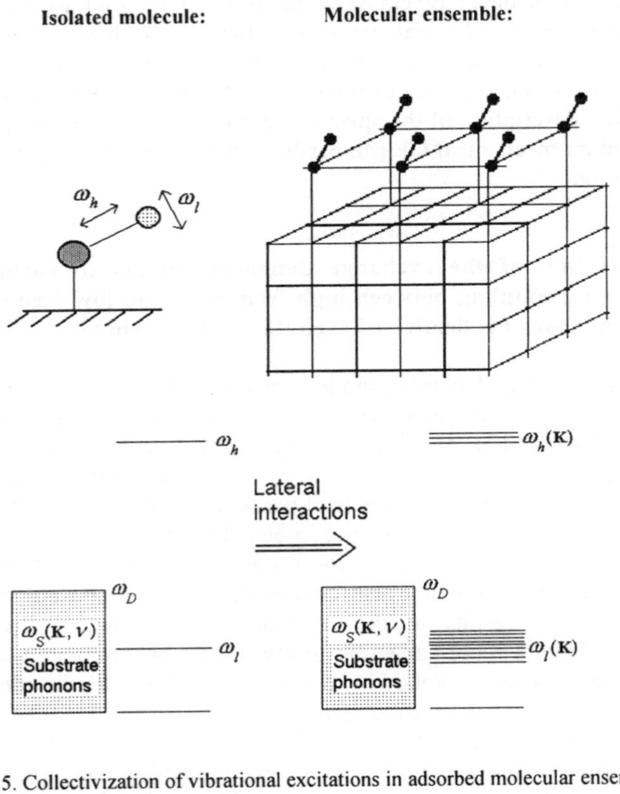

Fig. 4.5. Collectivization of vibrational excitations in adsorbed molecular ensemble due to intermolecular lateral interactions.

Considering only biquadratic anharmonic coupling, the dephasing of local vibrations was treated in the special case that only high-frequency modes underwent collectivization[174] and subsequently with the allowance also made for collectivized low-frequency modes.[138,175] It should be emphasized that the possibility for orientational phase transitions to occur in adsorbed molecular systems attaches special significance just to due regard for low-frequency mode collectivization, since dispersion laws for these modes are particularly sensitive to phase transition parameters. Nonetheless, these subtle effects, if included alone, without various anharmonic contributions, can hardly furnish the picture accounting for experimental evidence. The objective of the present section is to develop a consistent theory properly taking into account both lateral interactions of high-frequency and low-frequency modes of adsorbed molecules and all possible cubic and quartic anharmonic couplings between them.

Consider an arbitrary two-dimensional Bravais lattice, with its sites **R** occupied by adsorbed molecules and molecular vibrations representing two modes, of a high and low frequency. Frequencies $\omega_{h,l}$, reduced masses $m_{h,l}$, vibrational coordinates $u_{h,l}(\mathbf{R})$, and momenta $p_{h,l}(\mathbf{R})$ are accordingly labeled by subscripts h and l referring to the high-frequency and the low-frequency vibration. The most general form of the Hamiltonian appears as[140]

$$H^{(\mathrm{mol})} = H_h^{(\mathrm{mol})} + H_l^{(\mathrm{mol})} + H_{h-l}^{(\mathrm{mol})}, \tag{4.3.1}$$

$$\begin{aligned} H_{h,l}^{(\mathrm{mol})} &= \sum_{\mathbf{R}} \frac{p_{h,l}^2(\mathbf{R})}{2m_{h,l}} + \frac{1}{2}\sum_{\mathbf{R},\mathbf{R}'}\left[m_{h,l}\omega_{h,l}^2\delta_{\mathbf{R},\mathbf{R}'} + \Phi_{h,l,lat}(\mathbf{R}-\mathbf{R}')\right]u_{h,l}(\mathbf{R})u_{h,l}(\mathbf{R}') \\ &= \frac{1}{2}\sum_{\mathbf{K}}\left[\left|\tilde{p}_{h,l}(\mathbf{K})\right|^2 + \omega_{h,l}^2(\mathbf{K})\left|\tilde{u}_{h,l}(\mathbf{K})\right|^2\right], \end{aligned} \tag{4.3.2}$$

$$H_{h-l}^{(\mathrm{mol})} = V_3 + V_4, \tag{4.3.3}$$

$$V_3 = \sum_{\mathbf{R}}\left[\Phi_{30}u_h^3(\mathbf{R}) + \Phi_{21}u_h^2(\mathbf{R})u_l(\mathbf{R}) + \Phi_{12}u_h(\mathbf{R})u_l^2(\mathbf{R}) + \Phi_{03}u_\ell^3(\mathbf{R})\right], \tag{4.3.4}$$

$$\begin{aligned} V_4 = \sum_{\mathbf{R}}&\left[\Phi_{40}u_h^4(\mathbf{R}) + \Phi_{31}u_h^3(\mathbf{R})u_l(\mathbf{R}) + \Phi_{22}u_h^2(\mathbf{R})u_l^2(\mathbf{R})\right. \\ &\left. + \Phi_{13}u_h(\mathbf{R})u_l^3(\mathbf{R}) + \Phi_{04}u_\ell^4(\mathbf{R})\right]. \end{aligned} \tag{4.3.5}$$

The second equality in Eq. (4.3.2) demonstrates that the harmonic component of the Hamiltonian of the molecular subsystem is diagonalized by the Fourier transform in terms of wave vectors **K** of vibrational excitations:

$$u_{h,l}(\mathbf{R}) = (m_{h,l}N_0)^{-1/2} \sum_{\mathbf{K}} \tilde{u}_{h,l}(\mathbf{K}) e^{i\mathbf{K}\cdot\mathbf{R}},$$
$$p_{h,l}(\mathbf{R}) = (m_{h,l}/N_0)^{1/2} \sum_{\mathbf{K}} \tilde{u}_{h,l}(\mathbf{K}) e^{i\mathbf{K}\cdot\mathbf{R}}, \tag{4.3.6}$$

where N_0 is the number of adsorbate lattice sites in the main area. These collectivized excitations are characterized by the dispersion laws:

$$\omega_{h,l}^2(\mathbf{K}) = \omega_{h,l}^2 + \tilde{\Phi}_{h,l,\mathrm{lat}}(\mathbf{K})/m_{h,l}\,, \tag{4.3.7}$$

$$\tilde{\Phi}_{h,l,\mathrm{lat}}(\mathbf{K}) = \sum_{\mathbf{R}} \Phi_{h,l,\mathrm{lat}}(\mathbf{R}) e^{-i\mathbf{K}\cdot\mathbf{R}}. \tag{4.3.8}$$

In the particular case of dipole-dipole lateral interactions between molecules with the same dynamic dipole moment μ, we have:

$$\Phi_{h,l,\mathrm{lat}}(\mathbf{R}) = \sum_{\alpha,\beta=x,y,z} \left(\frac{\partial\mu_\alpha}{\partial u_{h,l}}\right)_0 \left(\frac{\partial\mu_\beta}{\partial u_{h,l}}\right)_0 D^{\alpha\beta}(\mathbf{R}), \tag{4.3.9}$$

$$D^{\alpha\beta}(\mathbf{R}) = \frac{\delta_{\alpha\beta}}{|\mathbf{R}|^3} - 3\frac{R_\alpha R_\beta}{|\mathbf{R}|^5} \tag{4.3.10}$$

(the Fourier components of the tensor $D^{\alpha\beta}(\mathbf{R})$ for various planar lattices were analyzed in detail in Sec. 2.2 and 3.2). Anharmonic coupling of high-frequency and low-frequency modes on the same lattice site is presented by relations (4.3.3)-(4.3.5).

The Hamiltonian of molecular vibrations (4.3.1), along with the Hamiltonian of substrate phonons (4.1.4) (which can also be represented as a sum over $\mathbf{K}$ and ν variables) and the operator of harmonic coupling between substrate phonons and low-frequency molecular modes (4.1.8), constitutes the full Hamiltonian of excitations in the adsorbate and the substrate. It represents a generalization for the above-considered Hamiltonian (4.2.12) in the following three respects. First, high-frequency and low-frequency molecular modes are collectivized as a result of lateral interactions in the adsorbate. Second, anharmonic coupling between high-frequency and low-frequency modes does not reduce to a biquadratic term and includes all possible cubic and quartic anharmonic contributions. Third, the generalized Hamiltonian is free of assumptions implied in the Heitler-London approximation. It is clear that with these generalizations, handling the problem exactly is no longer possible, but in practice it is often sufficient to apply the results provided by the fourth-order perturbation theory for two-time retarded Green's functions in the coordinate-momentum representation. Due to the occurrence of various types of anharmonic coupling and energy bands of molecular vibrations, the energy conservation law admits various processes contributing to the spectral line broadening for local vibrations (Table 4.1).

Table 4.1. Various processes contributing to the spectral line broadening for local vibrations. Frequencies of collectivized local vibrations $\Omega_{\mathbf{K}}$ (solid arrows) are supposed to exceed phonon frequencies $\omega_{\mathbf{Kq}}$ (dashed arrows): $\Omega_{\mathbf{K}} > \max \omega_{\mathbf{Kq}}$. For an extremely narrow band of local vibrations, diagrams A and B respectively refer to relaxation and dephasing processes, whereas diagrams C account for the case realizable only at the nonzero band width for local vibrations.

The number of concerted excitations	A	B	C
3		—	
4			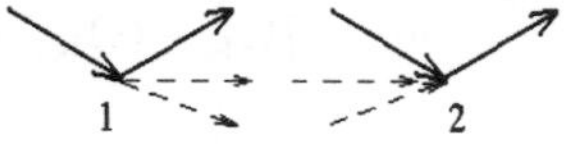

As the detailed mathematical description of these processes is rather tedious,[140] here we confine ourselves to the temperature-dependent dephasing and three-particle relaxation (processes of types B and 3-A in Table 4.1) contributing to the shift $\Delta_0(\Omega_{\mathbf{K}})$ and width $\Gamma_B(\Omega_{\mathbf{K}})$ of the spectral line for a local vibration $\Omega_{\mathbf{K}}$ at $\Omega_{\mathbf{K}} \equiv \omega_h(\mathbf{K}) >> \omega_{\mathbf{K}} \sim T$:

$$\Delta_0(\Omega_{\mathbf{K}}) = \frac{\hbar}{m_h \Omega_{\mathbf{K}} N_0} \sum_{\mathbf{K}_1} \frac{\Phi_{22}^{(\text{eff})}(\mathbf{K},\mathbf{K},\mathbf{K}_1)}{m_\ell \omega_{\mathbf{K}_1}} n(\omega_{\mathbf{K}_1}), \tag{4.3.11}$$

$$\Gamma_{\mathrm{B}}(\Omega_{\mathbf{K}})=\frac{\pi\hbar^2}{m_h^2 m_l^2 N_0^2}\frac{1}{}\sum_{\mathbf{K}_1\mathbf{K}_2}\frac{\left[\Phi_{22}^{(\mathrm{eff})}(\mathbf{K},\mathbf{K}_1,\mathbf{K}_2)\right]^2}{\Omega_{\mathbf{K}}\Omega_{\mathbf{K}_1}\omega_{\mathbf{K}_2}\omega_{\mathbf{K}-\mathbf{K}_1-\mathbf{K}_2}}n(\omega_{\mathbf{K}_2})\left[n(\omega_{\mathbf{K}-\mathbf{K}_1-\mathbf{K}_2})+1\right]$$
$$\times\Re\left(\Omega_{\mathbf{K}}-\Omega_{\mathbf{K}_1}+\omega_{\mathbf{K}_2}-\omega_{\mathbf{K}-\mathbf{K}_1-\mathbf{K}_2};\eta_{\mathbf{K}_2}+\eta_{\mathbf{K}-\mathbf{K}_1-\mathbf{K}_2}\right), \tag{4.3.12}$$

$$\Gamma_{\mathrm{A}}(\Omega_{\mathbf{K}})=\frac{\pi\hbar\Phi_{12}^2}{4m_h m_l^2\Omega_{\mathbf{K}}N_0}\sum_{\mathbf{K}_1}\frac{\Re\left(\Omega_{\mathbf{K}}-\omega_{\mathbf{K}_1}-\omega_{\mathbf{K}-\mathbf{K}_1};\eta_{\mathbf{K}_1}+\eta_{\mathbf{K}-\mathbf{K}_1}\right)}{\omega_{\mathbf{K}_1}\omega_{\mathbf{K}-\mathbf{K}_1}}\times\left[n(\omega_{\mathbf{K}_1})+n(\omega_{\mathbf{K}-\mathbf{K}_1})+1\right], \tag{4.3.13}$$

where the renormalized anharmonic coefficient can be represented as

$$\begin{aligned}\Phi_{22}^{(\mathrm{eff})}(\mathbf{K},\mathbf{K}_1,\mathbf{K}_2)&=\Phi_{22}\\&+3\frac{\Phi_{30}\Phi_{12}}{m_h}\frac{3\Omega_{\mathbf{K}}^2+\Omega_{\mathbf{K}_1}^2-\Omega_{\mathbf{K}-\mathbf{K}_1}^2}{[\Omega_{\mathbf{K}}^2-(\Omega_{\mathbf{K}_1}+\Omega_{\mathbf{K}-\mathbf{K}_1})^2][\Omega_{\mathbf{K}}^2-(\Omega_{\mathbf{K}_1}-\Omega_{\mathbf{K}-\mathbf{K}_1})^2]}\\&+3\frac{\Phi_{21}\Phi_{03}}{m_\ell}\frac{3\Omega_{\mathbf{K}}^2+\Omega_{\mathbf{K}_1}^2-\omega_{\mathbf{K}-\mathbf{K}_1}^2}{[\Omega_{\mathbf{K}}^2-(\Omega_{\mathbf{K}_1}+\omega_{\mathbf{K}-\mathbf{K}_1})^2][\Omega_{\mathbf{K}}^2-(\Omega_{\mathbf{K}_1}-\omega_{\mathbf{K}-\mathbf{K}_1})^2]}\\&+\sum_{\substack{\mathbf{K}_j=\mathbf{K}_2,\\ \mathbf{K}-\mathbf{K}_1-\mathbf{K}_2}}\left\{\frac{\Phi_{21}^2}{m_h}\frac{\Omega_{\mathbf{K}}^2+\omega_{\mathbf{K}_j}^2-\Omega_{\mathbf{K}-\mathbf{K}_j}^2}{[\Omega_{\mathbf{K}}^2-(\omega_{\mathbf{K}_j}+\Omega_{\mathbf{K}-\mathbf{K}_j})^2][\Omega_{\mathbf{K}}^2-(\omega_{\mathbf{K}_j}-\Omega_{\mathbf{K}-\mathbf{K}_j})^2]}\right.\\&\left.+\frac{\Phi_{12}^2}{m_\ell}\frac{\Omega_{\mathbf{K}}^2+\omega_{\mathbf{K}_j}^2-\omega_{\mathbf{K}-\mathbf{K}_j}^2}{[\Omega_{\mathbf{K}}^2-(\omega_{\mathbf{K}_j}+\omega_{\mathbf{K}-\mathbf{K}_j})^2][\Omega_{\mathbf{K}}^2-(\omega_{\mathbf{K}_j}-\omega_{\mathbf{K}-\mathbf{K}_j})^2]}\right\}.\end{aligned} \tag{4.3.14}$$

It is noteworthy that relation (4.3.12), with factors $(2\omega/\eta)\Re(\omega;\eta)$ substituted for $\Re(\omega;\eta)$, accounts for additional spectral line shifts $\Delta_1(\Omega_{\mathbf{K}})$ for local vibrations which arise from the corresponding process. However, this contribution does not comprise the total line shift in this high-order perturbation theory just as contribution (4.3.11) does not provide total shift in the second-order perturbation theory.

It is expedient first to analyze the contribution from cubic anharmonicity in the simplest case, i.e. for a system free of lateral interactions, when $\Omega_{\mathbf{K}}$ and $\omega_{\mathbf{K}}$ are independent of **K** and amount respectively to $\Omega_0 = \omega_h$ and $\omega_0 = \omega_l$. With regard for the inequalities $\eta_0 << \omega_0 < \Omega_0$, relations (4.3.11)-(4.3.13) yield:

$$\Delta(\Omega_0) = \gamma_0^{(\text{eff})} n(\omega_0), \tag{4.3.15}$$

$$\Gamma_B(\Omega_0) = \frac{\left[\gamma_0^{(\text{eff})}\right]^2}{\eta_0} n(\omega_0)\left[n(\omega_0)+1\right], \tag{4.3.16}$$

$$\Gamma_A(\Omega_0) = \frac{\hbar \Phi_{12}^2}{2m_h m_l^2 \Omega_0 \omega_0^2}\left[n(\omega_0)+\frac{1}{2}\right]\frac{\eta_0}{(\Omega_0 - 2\omega_0)^2 + \eta_0^2}, \tag{4.3.17}$$

$$\gamma_0^{(\text{eff})} = \frac{\hbar}{m_h m_l \Omega_0 \omega_0}\left\{\Phi_{22} - 3\frac{\Phi_{30}\Phi_{12}}{m_h \Omega_0^2} - \frac{2\Phi_{21}^2}{m_h(4\Omega_0^2 - \omega_0^2)} - 3\frac{\Phi_{21}\Phi_{03}}{m_\ell \omega_0^2} + \frac{\Phi_{12}^2}{2m_l\omega_0}\left[\frac{\Omega_0 - 2\omega_0}{(\Omega_0 - 2\omega_0)^2 + \eta_0^2} - \frac{1}{\Omega_0 + 2\omega_0}\right]\right\}. \tag{4.3.18}$$

The theory developed permits spectral line shift and width to be calculated from Taylor power series for interatomic potential energies in a concrete system. Various methods of tackling this problem can be found in the literature[140,169,171,176-180] (see also survey 181 and references cited therein). Here we invoke a realistic model for the coupling of two mutually perpendicular vibrations which was reported by Burke, Langreth, Persson, and Zhang.[1] As in Ref. 1, write the Hamiltonian for the interaction between the modes u_h and u_l in polar coordinates r and θ, where θ is the angle between the adsorbate bond and the perpendicular to the surface plane:

$$H^{(\text{mol})} = \frac{p_r^2}{2m} + \frac{p_\theta^2}{2m(r_0 + u_r)^2} + \frac{1}{2}k_\theta \theta^2 + \frac{1}{2}k_\theta u_r^2 + \Phi_3 u_r^3 + \ldots \tag{4.3.19}$$

(r_0 is the equilibrium length of the adsorbate-surface bond). Expanding the factor $(r_0+u_r)^{-2}$ in u_r and taking into account that $p_\theta^2 / m r_0^2$ has the same amplitude as $k_\theta \theta^2$, we are led to the following values of the parameters involved:

$$\Omega_0 = (k_r/m)^{1/2}, \quad \omega_0 = \left(k_\theta / m r_0^2\right)^{1/2}, \quad u_\ell = r_0\theta, \quad \Phi_{30} = \Phi_3, \qquad (4.3.20)$$
$$\Phi_{21} = \Phi_{03} = 0, \quad \Phi_{12} = -m\omega_0^2/r_0, \quad \Phi_{22} = 3m\omega_0^2/2r_0^2 .$$

(In contrast to Eq. (3.13) of Ref. 1 where $u_l=(r_0+u_r)\sin\theta$, we assume that the force constant k_θ of the frustrated rotation with the changing angle θ is independent of u_r). On substituting values (4.3.20) in Eq. (4.3.18) and accounting for the doubled contribution from two modes ω_0 mutually perpendicular in the surface plane, we obtain:

$$\gamma_0^{(\text{eff})} = \frac{3\hbar\omega_0}{m r_0^2 \Omega_0}\left(1 + \frac{2\Phi_3 r_0}{k_r} + \delta\right), \qquad (4.3.21)$$

$$\delta = \frac{1}{3}\omega_0\left[\frac{\Omega_0 - 2\omega_0}{(\Omega_0 - 2\omega_0)^2 + \eta^2} - \frac{1}{\Omega_0 + 2\omega_0}\right]\Bigg\}, \qquad (4.3.22)$$

$$\Gamma_A^{(3)}(\Omega_0) = \frac{\hbar\omega_0^2}{m r_0^2 \Omega_0}\left[n(\omega_0) + \frac{1}{2}\right]\frac{\eta}{(\Omega_0 - 2\omega_0)^2 + \eta^2} . \qquad (4.3.23)$$

Evidently, we would arrive at the same result if the problem were considered in the Cartesian system of coordinates, with accordingly different values of anharmonic coupling coefficients:

$$\Phi_{12} = m\left(\Omega_0^2 - 2\omega_0^2\right)/2r_0, \Phi_{22} = -\left(m\Omega_0^2 - 3m\omega_0^2 - 3\Phi_3 r_0\right)/2r_0^2 \qquad (4.3.24)$$

containing contributions from central forces (proportional to Ω_0^2 and Φ_3). However, it follows from the structure of the relationship (4.3.18) that these contributions to $\gamma_0^{(\text{eff})}$ should be completely compensated, just as in the limit $\Omega_0 >> \omega_0$ treated in Ref. 1. To derive Eq. (4.3.23) in Cartesian coordinates, we should invoke the energy conservation law, $\Omega_0=2\omega_0$, before using Sokhotskii's formula, $(x + i0)^{-1} = P\left(x^{-1}\right) - i\pi\delta(x)$, which introduces vibration relaxation in the system concerned. Thus the model specified by Eq. (4.3.19) serves for the verification of relations (4.3.21)-(4.3.23) gained in the high-order perturbation theory.

For the case of CO/Pt(111), Persson and Ryberg[176] regarded a parallel frustrated CO translation as a low-frequency parallel mode and derived the estimate

$$\gamma^{(\mathrm{Pers})} = -\hbar\,\omega_0\Omega_0/4E_h \tag{4.3.25}$$

which was also in nice accordance with experimental data for the systems H/Si(111)[182] and H/C(111).[171] To arrive at Eq. (4.3.25), they in fact introduced a new Morse potential which accounted for potential energies of two mutually perpendicular displacements u_h and u_l, with the u_l-dependence included only in the E_h parameter, so that the free term in the expansion of this potential in u_h specified the harmonic potential energy $m\omega_0^2 u_\ell^2/2$ of the vibration u_l:[176]

$$E^{(\mathrm{Pers})}(u_h,u_\ell) = \left(E_h - \frac{1}{2}m\omega_0^2 u_\ell^2\right)\left(e^{-2\alpha u_h} - 2e^{-\alpha u_h}\right), \quad \alpha = \sqrt{\frac{m\Omega_0^2}{2E_h}}. \tag{4.3.26}$$

(Here E_h and α are the parameters of a normal Morse potential, and $m \equiv m_h = m_l$ in the framework of the model discussed). Although this combined potential (4.3.26) provides plausible estimates, it can hardly be substantiated in terms of the theory of intermolecular interactions and contains, in addition, only biquadratic anharmonic coupling $-(m\omega_0^2\alpha^2/2)u_h^2u_\ell^2$ (leading to Eq. (4.3.25)); thus, it cannot be involved as an intermolecular potential in the approach defined by relations (4.3.21)-(4.3.23) which also takes account of the cubic anharmonic coupling.

Of interest is to compare Eqs. (4.3.21) and (4.3.25); for this purpose, express k_r and Φ_3 in terms of the parameters E_h and α of the normal Morse potential and introduce the ratio of the equilibrium adsorbate-surface bond length to α^{-1}: $\kappa = \alpha r_0$. Then $B=-\gamma^{(\mathrm{Pers})}\Omega_0/\kappa^2\omega_0$, $2\Phi_3 r_0/k_r = -\kappa$ and we are led to

$$\gamma_0^{(\mathrm{eff})} = \frac{6(\kappa-1-\delta)}{\kappa^2}\gamma^{(\mathrm{Pers})}. \tag{4.3.27}$$

The parameter values κ and δ for light adatoms are listed in Table 4.2. The proportionality factor for $\gamma_0^{(\mathrm{eff})}$ and $\gamma^{(\mathrm{Pers})}$ (see Eq. (4.3.27)) proves to be of the order of unity, which makes our estimates close to those by Persson and Ryberg.[176] Thus, it is hardly surprising that the results of the approach presented in Eqs. (4.3.21)-(4.3.23) are consistent with the experimental evidence. The specificity of this model becomes noticeable if the value Ω_0 is close to $2\omega_0$ and hence the resonance factor δ grows large enough. For the system H/C(111), positive values of δ somewhat decrease the absolute value of $\gamma_0^{(\mathrm{eff})}$, while the system D/C(111) is characterized by $\delta<0$ and accordingly larger absolute value of $\gamma_0^{(\mathrm{eff})}$. As the absolute value of the parameter δ increases, so does the contribution of the relaxation process $\Gamma_A^{(3)}$ which becomes predominant for the systems H(D)/C(111).

Table 4.2. The basic parameters of the models under discussion for some adsorption systems. The dephasing ($2\Gamma_B^{(4)}$) and the relaxation ($2\Gamma_A^{(3)}$) contributions to the full spectral linewidth for local vibrations as well as the experimentally observed values of this parameter ($2\Gamma^{(exp)}$) are presented for the temperature T=300 K. Data are taken from a, Ref. 176; b, Ref. 169.

Parameter	H/Si(111)	H/C(111)	D/C(111)
Ω_0 (cm^{-1})	2086	2835	2110
E_h (eV)	3.5	3.5	3.5
ω_0 (cm^{-1})	210	1140	1140
κ	2.21	2.17	2.20
η (cm^{-1})	52	120	120
δ	0.014	0.579	-1.58
$\gamma^{(Pers)}$ (cm^{-1})	-3.88	-28.6	-21.3
$\gamma_0^{(eff)}$ (cm^{-1})	-5.7	-21.6	-73.3
$\gamma_0^{(exp)}$ (cm^{-1})	-5[a]	-23[b]	-
$2\Gamma_B^{(4)}$ (cm^{-1})	1.13	0.03	0.38
$2\Gamma_A^{(3)}$ (cm^{-1})	0.01	5.19	28.0
$2\Gamma^{(exp)}$ (cm^{-1})	1[a]	5.7[b]	30[b]

4.3.1. Contribution of dipolar dispersion laws to dephasing of high-frequency collective vibrations

Here we focus on the effect of dipolar dispersion laws for high-frequency collective vibrations on the shift and width of their spectral line, with surface molecules inclined at an arbitrary angle θ to the surface-normal direction. For definiteness, we consider the case of a triangular lattice and the ferroelectric ordering of dipole moments inherent in this lattice type.[56,109] Lateral interactions of dynamic dipole moments $\boldsymbol{\mu} = \mu\,\mathbf{e}$ ($\mathbf{e} = (\sin\theta\cos\varphi, \sin\theta\sin\varphi, \cos\theta)$) corresponding to collective vibrations on a simple two-dimensional lattice of adsorbed molecules cause these vibrations to collectivize in accordance with the dispersion law:[121]

$$\Omega_{\mathbf{k}}^2 = \Omega_h^2 + \frac{1}{m_h a^3} \sum_{\alpha,\beta=x,y,z} \left(\frac{\partial \mu_\alpha}{\partial u_h}\right)_0 \left(\frac{\partial \mu_\beta}{\partial u_h}\right)_0 \widetilde{D}^{\alpha\beta}(\mathbf{k}). \tag{4.3.28}$$

Here

$$\widetilde{D}^{\alpha\beta}(\mathbf{k}) = \sum_{\mathbf{R}} D^{\alpha\beta}(\mathbf{R}) e^{-i\mathbf{k}\cdot\mathbf{R}} \tag{4.3.29}$$

is the Fourier component of the tensor of dipole-dipole interactions (4.3.10), R is measured in units of the lattice constant a; m_h and u_h are the reduced mass and the displacement of the vibration with the frequency Ω_h. Since the spectral lines observed in the IR region are determined by the value $\mathbf{k}=0$ at which the tensor $\widetilde{D}^{\alpha\beta}(0) = \widetilde{D}(0)\delta_{\alpha\beta}$ is isotropic for symmetric lattices, the corresponding spectral shift caused by dipole-dipole interactions can be written as:

$$\Delta\Omega_0 \equiv \Omega_0 - \Omega_h \approx \Omega_{\text{dip}} \widetilde{D}(0), \quad \Omega_{\text{dip}} = \frac{\chi_{\text{v}}}{2a^3}\Omega_h, \quad \chi_{\text{v}} = \frac{1}{m_h}\left(\frac{\partial \mu}{\partial u_h}\right)_0^2,$$
$$\widetilde{D}(0) = \frac{1}{2}\left(3\cos^2\theta - 1\right)D_0, \quad D_0 \approx 11.034, \tag{4.3.30}$$

where χ_{v} is the vibrational polarizability of the molecule. Thus, the positive and negative shifts result at $\theta < 54.7^0$ and $\theta > 54.7^0$, respectively. The approximate equality in Eq. (4.3.30) holds at $|\Delta\Omega_0| \ll \Omega_h$, which is mostly the case for real systems. With the electronic polarizability taken into account, the value Ω_{dip} is renormalized by the factor $(1 + \chi_{\text{e}}\widetilde{D}(0)/a^3)^{-1}$.[183] The parameter Ω_{dip} can also be influenced by image dipole effects which are sometimes quite noticeable even on insulator surfaces.

An additional shift and broadening of the spectral line for high-frequency collective vibrations in the framework of the widespread dephasing model[147,148,151] are describable in terms of the anharmonic coupling of these vibrations with a low-frequency resonance mode ω_l characterized by finite lifetimes $\tau = 2/\eta$ (η is the full resonance width). With the anharmonic coupling specified by the function $\gamma_{\mathbf{k}}$, the maximum position $\Omega_{\max}$ and the width Γ of the spectral line at $|\gamma_{\mathbf{k}}| \ll \eta + |\Delta\Omega_0|$ are defined by the following relationships:

$$\Omega_{\max} \equiv \Omega_0 + \gamma_0/2 + \Delta, \quad \Delta = \Delta_0 + \Delta_1, \quad \Delta_0 = \gamma_0 n(\omega_l) \tag{4.3.31}$$

$$\begin{pmatrix} \Delta_1 \\ \Gamma \end{pmatrix} = n(\omega_l)[n(\omega_l)+1]\frac{1}{N}\sum_{\mathbf{k}} \frac{|\gamma_{\mathbf{k}}|^2}{(\Omega_0 - \Omega_{\mathbf{k}})^2 + \eta^2} \begin{pmatrix} \Omega_0 - \Omega_{\mathbf{k}} \\ \eta \end{pmatrix} \tag{4.3.32}$$

where N is the number of adsorbate lattice sites in the main area, and

$$n(\omega) = [\exp(\hbar\omega/k_B T) - 1]^{-1} \tag{4.3.33}$$

is the temperature-dependent factor of the Bose-Einstein statistics. Persson, Hoffman, and Ryberg were the first to derive an expression like Eq. (4.3.32) which involved the coefficient γ independent of the wave vector **k** and concerned the case of biquadratic anharmonic coupling between the collective and the resonance mode.[174] For the same type of the anharmonic coupling considered in the low-temperature limit ($k_B T << \hbar\omega_l$, γ, η, and $\Delta\Omega_0$ are arbitrary), the following expression is valid:[184]

$$\begin{pmatrix} \Delta \\ \Gamma \end{pmatrix} = \begin{pmatrix} \mathrm{Re} \\ -\mathrm{Im} \end{pmatrix} W, \quad W = \gamma n(\omega_l)\left[1 - \frac{\gamma}{N}\sum_{\mathbf{k}} \frac{1}{\Omega_0 - \Omega_{\mathbf{k}} + i\eta}\right]^{-1}. \tag{4.3.34}$$

(This formula is derived in Appendix 3). With regard to various cubic and quartic anharmonic interactions, the quantity $\gamma_{\mathbf{k}}$ is characterized by a certain combination of these anharmonic contributions and becomes dependent on **k** (see Eq. (4.3.14) for a related quantity and Ref. 140). However, this dependence is insignificant compared to the **k**-dependence appearing in the denominators of Eqs. (4.3.32) and (4.3.34). Therefore, spectral characteristics defined by formulae (4.3.32) can with good reason be regarded as proportional to certain functions of lateral interaction parameters and of the resonance width η:

$$\begin{pmatrix} F_\Delta \\ F_\Gamma \end{pmatrix} = \begin{pmatrix} \mathrm{Re} \\ -\mathrm{Im} \end{pmatrix} F, \quad F \equiv \frac{1}{N}\sum_{\mathbf{k}} \frac{1}{\Omega_0 - \Omega_{\mathbf{k}} + i\eta} = \int_{-\infty}^{\infty} \frac{\rho_h(\Omega)d\Omega}{\Omega_0 - \Omega + i\eta} \tag{4.3.35}$$

where

$$\rho_h(\Omega) = \frac{1}{N}\sum_{\mathbf{k}} \delta(\Omega - \Omega_{\mathbf{k}}) \tag{4.3.36}$$

is the spectral density function for collective vibrations. On the other hand, the same functions enter in the low-temperature approximation (see Eq. (4.3.34)):

$$\begin{pmatrix} \Delta \\ \Gamma \end{pmatrix} = \frac{\gamma n(\omega_l)}{(1-\gamma F_\Delta)^2 + \gamma^2 F_\Gamma^2} \begin{pmatrix} 1-\gamma F_\Delta \\ \gamma F_\Gamma \end{pmatrix}. \tag{4.3.37}$$

For functions (4.3.35) to be calculated, rather complicated specific dispersion laws should be known for the excitations induced by long-range anisotropic dipole forces. That is why, the approximate expression

$$F^{(\mathrm{appr})} = (\Delta\Omega_0 + i\eta)^{-1} \tag{4.3.38}$$

is frequently used in estimations. Here all characteristics of lateral interactions are accounted for by a single parameter $\Delta\Omega_0$ defined in Eq. (4.3.30). Substituting Eq. (4.3.38) in (4.3.37), we arrive at the known low-temperature formula of Erley and Persson:[185]

$$\begin{pmatrix} \Delta_1^{(\mathrm{appr})} \\ \Gamma^{(\mathrm{appr})} \end{pmatrix} = \frac{\gamma^2 n(\omega_l)}{(\Delta\Omega_0 - \gamma)^2 + \eta^2} \begin{pmatrix} \Delta\Omega_0 - \gamma \\ \eta \end{pmatrix}. \tag{4.3.39}$$

In what follows we will establish under which conditions and to which accuracy the approximation (4.3.38) is justified for the cases of normal, parallel, and inclined molecular orientations with reference to the surface plane.[186]

The curves 1 in Figs. 4.6a and b show the functions F_Γ and F_Δ calculated by formulae (4.3.35) and (4.3.38) for the case of normal molecular orientations ($\mathbf{e} \parallel Oz$) and plotted versus the argument $\Delta\Omega/(\eta+\Delta\Omega)$. The dimensionless argument and functions of this kind normalized with respect to the sum of the resonance and the band widths were introduced so as to depict their behavior in both limiting cases, $\Delta\Omega \ll \eta$ and $\Delta\Omega \gg \eta$. The deviation of the solid lines from the dotted ones indicates to which degree the one-parameter approximation defined by Eq. (4.3.38) differs from the realistic dispersion law. As seen, this approximation shows excellent adequacy, but for the region $\Delta\Omega \gg \eta$, where the asymptotic behavior of the approximation (4.3.38) and Eq. (4.3.35) are as follows:

$$F_{\Delta}^{(\mathrm{appr})} \approx \frac{1.2}{\Delta\Omega}\left(1 - 1.44\frac{\eta^2}{\Delta\Omega^2}\right),\ F_{\Gamma}^{(\mathrm{appr})} \approx 1.44\frac{\eta}{\Delta\Omega^2}, \tag{4.3.40}$$

$$F_{\Delta} \approx \frac{1}{\Delta\Omega}\left(2\ln 2 - \frac{\pi}{4}\frac{\eta}{\Delta\Omega}\right),\ F_{\Gamma} \approx \frac{\eta}{2\Delta\Omega^2}\ln\frac{4e\Delta\Omega}{\eta}. \tag{4.3.41}$$

As the behavior of the spectral line width, $\eta\ln(1/\eta)$, can roughly be regarded as linear in η and the coefficient value $2\ln 2 \approx 1.386$ is close to 1.2, the approximation (4.3.38) holds workable for normal orientations even at $\Delta\Omega >> \eta$.

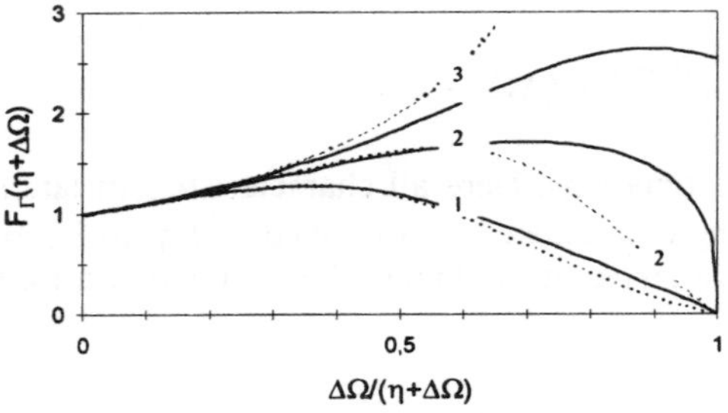

a

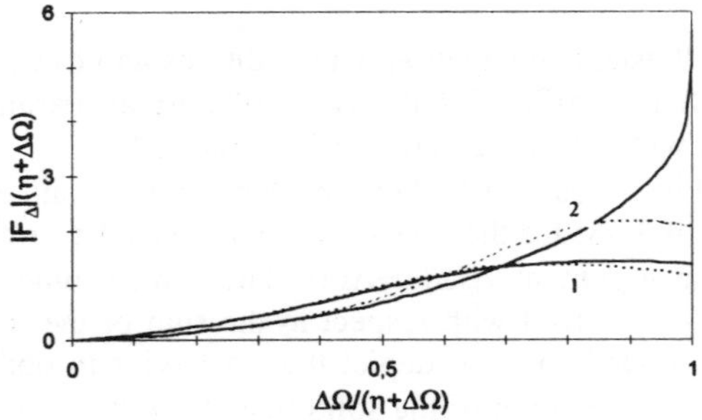

b

Fig. 4.6. Dependences of dimensionless spectral characteristics $F_{\Gamma}(\eta+\Delta\Omega)$ (a) and $F_{\Delta}(\eta+\Delta\Omega)$ (b) (as defined in Eq. (4.3.35)) versus the dimensionless parameter $\Delta\Omega/(\eta+\Delta\Omega)$ at $\theta = 0$, $\Delta\Omega = 13.241\Omega_{\mathrm{dip}}$ (1); $\theta = 90^0$, $\Delta\Omega = 11.586\Omega_{\mathrm{dip}}$ (2); $\theta = 54.7^0$, $\Delta\Omega = 6.080\Omega_{\mathrm{dip}}$ (3). Solid and dotted curves correspond to Eq. (4.3.35) and to the approximation specified by Eq (4.3.38).

The case of parallel orientations ($\mathbf{e} \perp Oz$) differs radically from the previously considered one, since the frequency dependence of the spectral density function is specified by the fractional power law:[109]

$$\rho_h(\Omega) = \frac{(3/2\pi)^{1/2}}{\Gamma^2(1/4)\gamma_\parallel^{1/2}\gamma_\perp^{3/4}} \frac{1}{\Omega_{\mathrm{dip}}} \left(\frac{\Omega - \Omega_0}{\Omega_{\mathrm{dip}}} \right)^{1/4},$$

$$\Omega \geq \Omega_0, \quad \gamma_\parallel = \frac{4\pi}{\sqrt{3}} \approx 7.255, \quad \gamma_\perp \approx 0.263 . \tag{4.3.42}$$

(Here $\Gamma(1/4) \approx 3.626$ is the special value of the gamma function so that the numerical value of the coefficient is 0.0531). On substituting Eq. (4.3.42) in (4.3.35), we obtain, at $\eta \ll \Delta\Omega$, the following relationships:

$$F_\Gamma = \frac{\pi}{\sqrt{2+\sqrt{2}}} \frac{(3/2\pi)^{1/2}}{\Gamma^2(1/4)\gamma_\parallel^{1/2}\gamma_\perp^{3/4}} \frac{1}{\Omega_{\mathrm{dip}}} \left(\frac{\eta}{\Omega_{\mathrm{dip}}} \right)^{1/4} . \tag{4.3.43}$$

With $\eta \ll \Delta\Omega$, the distinction between solid and dotted lines 2 in Fig. 4.5 becomes significant suggesting that the approximation defined in Eq. (4.3.38) is inapplicable to the case of parallel orientations.

The angular dependences $F_\Gamma\Delta\Omega$ and $F_\Delta\Delta\Omega$ at $\eta \to 0$ calculated for inclined molecular orientations are represented by the curves 1 and 2 in Fig. 4.7. The spectral density function (appearing as F_Γ/π at $\eta \to 0$) takes nonzero values if the angle θ ranges from 46.7^0 to 90^0, since in this θ range the value Ω_0 is neither the upper (as at $\theta < 46.7^0$) nor the lower ($\theta = 90^0$) edge of the collective vibration band. The maximum values F_Γ found in the vicinity of the angles $\theta = 49^0$ and $\theta = 63^0$ correspond to the regions of dispersion laws with small values $|\partial\Omega_{\mathbf{k}}/\partial\mathbf{k}|$. The asymptotically exact result with a nonzero spectral density function can be obtained for molecules which are only slightly inclined to the surface plane:

$$\rho_h(\Omega_0) = \frac{\sqrt{3}}{2\pi^2\gamma_\perp} \frac{\cos\theta}{\Omega_{\mathrm{dip}}}, \quad \cos\theta \ll 1 . \tag{4.3.44}$$

For the value F_Δ to be calculated, the integration over the whole Brillouin zone should be performed which yields, in the limiting case of $\theta = 90^0$, the value $F_\Delta \approx -5.44/\Omega_{\mathrm{dip}}$ (in the framework of the isotropic approximation).

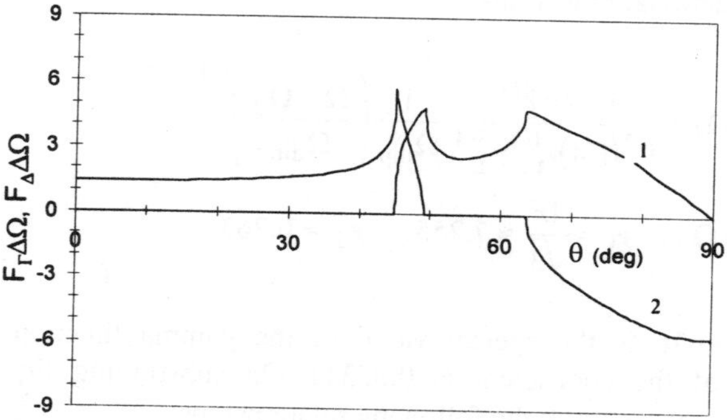

Fig. 4.7. Angular dependences of dimensionless spectral characteristics $F_\Gamma\Delta\Omega$ (curve 1) and $F_\Delta\Delta\Omega$ (curve 2) calculated at $\eta = 0$.

As a main point of this subsection,[186] it is shown that due to sufficiently strong lateral interactions of adsorbed molecules on a two-dimensional triangular lattice, the spectral line for collective vibrations manifests some characteristic peculiar relationships between its dephasing-induced broadening and the resonance width η for the low-frequency mode: The line width changes as $\eta\ln(1/\eta)$ for surface-normal and as $\eta^{1/4}$ for surface-parallel molecular orientations, and takes nonzero values (independent of η) for inclined molecules with the inclination angle ranging from 47^0 to 90^0.

4.3.2. A simple model for collective high-frequency and low-frequency molecular modes

In the framework of the low-temperature dephasing model, the dependence of spectral line shifts and widths for local vibrations on real dispersion laws for high-frequency ($\Omega_\mathbf{k}$) and low-frequency ($\omega_\mathbf{k}$) molecular modes is defined by the complex function (see Appendix 3):[184]

$$W_\mathbf{k} = \frac{\gamma}{N_0} \sum_{\mathbf{k}_1} \frac{n(\omega_{\mathbf{k}_1})}{1 - \gamma F_\mathbf{k}(\mathbf{k}_1)}, \tag{4.3.45}$$

$$F_{\mathbf{k}}(\mathbf{k}_1)=\frac{1}{N_0}\sum_{\mathbf{k}_2}\left[\Omega_{\mathbf{k}}-\Omega_{\mathbf{k}+\mathbf{k}_2-\mathbf{k}_1}+\omega_{\mathbf{k}_1}-\omega_{\mathbf{k}_2}+i\left(\eta_{\mathbf{k}_1}+\eta_{\mathbf{k}_2}\right)/2\right]^{-1}. \quad (4.3.46)$$

Here **k**, $\mathbf{k}_1$, $\mathbf{k}_2$ are dimensionless wave vectors of vibrational excitations in a lattice of adsorbed molecules (with N_0 molecules in the main area).

Formulae of Erley and Persson[185] (4.3.38) and (4.3.39) follow from general relationships (4.3.45) and (4.3.46) with $\mathbf{k} = 0$ provided the dispersion of low-frequency modes, $\omega_{\mathbf{k}}$ and $\eta_{\mathbf{k}}$, is neglected and the differences $\Omega_{\mathbf{k}}-\Omega_{\mathbf{k}+\mathbf{k}_2-\mathbf{k}_1}$ are substituted by the corresponding values of spectral shifts $\Delta\Omega_{\mathbf{k}}$. To estimate the effect caused by the dispersion of low-frequency molecular modes on the local vibration spectrum, we substitute the difference $\omega_{\mathbf{k}_1}-\omega_{\mathbf{k}_2}$ in Eq. (4.3.46) by the parameter $\Delta\omega_{\mathbf{k}}$, the effective halfwidth of the band of low-frequency vibrations corresponding to the high-frequency spectral line with the wave vector **k**. With regard to the sign change resulting from the permutation of the summation variables $\mathbf{k}_1$ and $\mathbf{k}_2$ in this difference, we arrive at the following generalization[52,187] of the Erley-Persson formulae:

$$W_{\mathbf{k}}=\frac{\gamma n(\omega_{\mathbf{k}})}{1-\gamma F_{\mathbf{k}}}, \quad F_{\mathbf{k}}=\frac{\Delta\Omega_{\mathbf{k}}+i\eta}{(\Delta\Omega_{\mathbf{k}}+i\eta)^2-\Delta\omega_{\mathbf{k}}^2}. \quad (4.3.47)$$

Owing to the additional parameter $\Delta\omega_{\mathbf{k}}$ that enters into the above expression, the sign of $\mathrm{Re}F_{\mathbf{k}}$ is opposite to that of $\Delta\Omega_{\mathbf{k}}$ at $\Delta\omega_{\mathbf{k}}^2 > \Delta\Omega_{\mathbf{k}}^2 + \eta^2$. This fact proves to be fundamentally important in the analysis of temperature dependences of the spectral line shifts for the system CO/NaCl(100).

Consider a 2x1 phase of the monolayer of CO molecules forming a square lattice on the NaCl(100) surface. The orientational inequivalence of two CO molecules in the unit cell of the two-dimensional lattice (see Fig. 2.3) results in Davydov-split spectral lines for radial C-O vibrations with the frequencies Ω_S = 2154.86 and Ω_A = 2148.58 cm^{-1} at T = 11 K (see Fig. 2.5) that correspond to symmetric surface-normal (S) and antisymmetric surface-parallel (A) normal vibrations.[99] The dispersion law for these vibrations, if calculated with regard to geometrical features of the system (molecular orientations inclined at an angle of 25^0 to the surface normal with alternating azimuthal angles, $\varphi = 0$ and $\varphi = \pi$, along any of the square lattice axes), reproduces the observed magnitude of the Davydov splitting, $\Omega_S - \Omega_A \approx 6.3$ cm^{-1}, and yields the following values for spectral shifts: $\Delta\Omega_S \approx 5.1$, $\Delta\Omega_A \approx -1.2$ cm^{-1}.[81] Temperature-dependent contributions to shifts and widths of split spectral lines in the temperature range from 11 to 20 K are presented in Fig. 4.8 and can be approximated as follows:

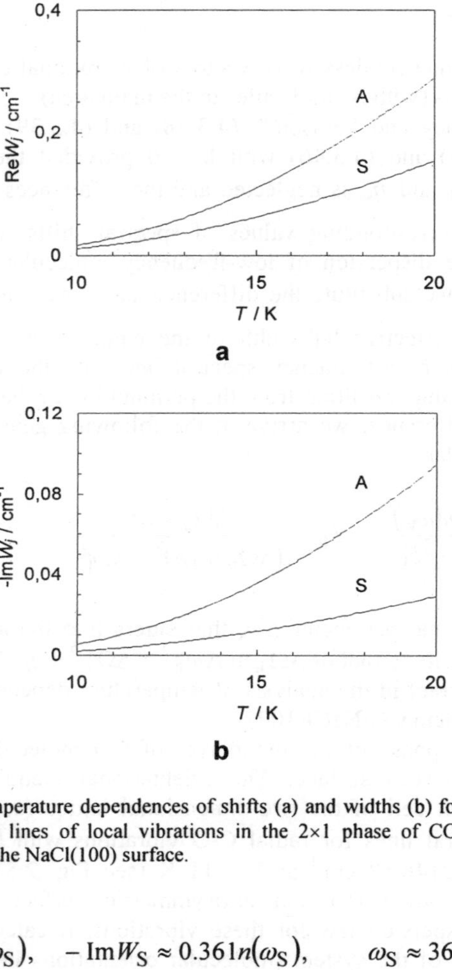

Fig. 4.8. Temperature dependences of shifts (a) and widths (b) for Davydov-split spectral lines of local vibrations in the 2×1 phase of CO molecules adsorbed on the NaCl(100) surface.

$$\begin{aligned}
&\mathrm{Re}W_{\mathrm{S}} \approx 1.90n(\omega_{\mathrm{S}}), \quad -\mathrm{Im}W_{\mathrm{S}} \approx 0.361n(\omega_{\mathrm{S}}), \qquad \omega_{\mathrm{S}} \approx 36\mathrm{cm}^{-1}, \\
&\mathrm{Re}W_{\mathrm{A}} \approx 4.50n(\omega_{\mathrm{A}}), \quad -\mathrm{Im}W_{\mathrm{A}} \approx 1.473n(\omega_{\mathrm{A}}), \qquad \omega_{\mathrm{A}} \approx 39\mathrm{cm}^{-1}.
\end{aligned} \tag{4.3.48}$$

The nature of these dependences changes dramatically at T > 20 K, as an orientational transition from the 2×1 to the 1×1 phase of CO molecules results at T ≈ 20 K.

To describe the temperature dependences observed, one can first attempt to take advantage of well-known formulae by Erley and Persson[185] (4.3.39) written for symmetric (j=S) and antisymmetric (j=A) vibrations:

$$\mathrm{Re}W_j = \gamma n(\omega_j)\left[1+\gamma\frac{\Delta\Omega_j-\gamma}{(\Delta\Omega_j-\gamma)^2+\eta^2}\right],$$
$$-\mathrm{Im}W_j = \gamma^2 n(\omega_j)\frac{\eta}{(\Delta\Omega_j-\gamma)^2+\eta^2}. \tag{4.3.49}$$

Since it follows from Eq. (4.3.48) that $|\mathrm{Im}W_S| < |\mathrm{Im}W_A|$, the second formula in Eq. (4.3.49) leads to the inequality $|\Delta\Omega_S - \gamma| > |\Delta\Omega_A - \gamma|$. To account for positive shifts, it is necessary that $\gamma > 0$. Moreover, $\Delta\Omega_S > 0$ and $\Delta\Omega_A < 0$. However, with this conditions included, the first formula in Eq. (4.3.49) can yield only $\mathrm{Re}W_S > \mathrm{Re}W_A$, whereas just the reverse is true for real systems.

To determine the characteristics of the 2×1 phase in the system CO/NaCl(100) from general formulae (4.3.47), we equate expressions (4.3.47) and (4.3.48) thus deriving four equations in four unknown parameters, γ, η and $\Delta\omega_j$ with j = S, and A. It is noteworthy that for the spectral lines associated with local vibrations S and A, the vector **k** assumes two values: $\mathbf{k} = 0$ and $\mathbf{k} = \mathbf{k}_A$ ($\mathbf{k}_A$ is a symmetric point at the boundary of the first Brillouin zone). The exact solution of the system of equations provides parameter values listed in Table 4.3.[187] The same parameters were previously evaluated by formulae (4.3.49) without regard for lateral interactions of low-frequency molecular modes.[99] As a consequence, the result was physically meaningless: the quantities γ and η proved to be different for vibrations S and A (also see Table 4.3).

Table 4.3. Parameters of Davydov-split spectral lines for the 2×1 phase of the system CO/NaCl(100) calculated by the formulae of Erley and Persson (4.3.49) and by the generalized formula (4.3.47) accounting for the dispersion of low-frequency CO vibrations.

	Values calculated by Eqs. (4.3.49) (cm^{-1})		Values calculated by Eqs. (4.3.47) (cm^{-1})	
Parameters	Surface-normal symmetric vibrations (j=S)	Surface-parallel antisymmetric vibrations (j=A)	Surface-normal symmetric vibrations (j=S)	Surface-parallel antisymmetric vibrations (j=A)
$\Delta\Omega_j$	4.6	-1.9	5.1	-1.2
$\Delta\omega_j$	-	-	6.8	3.6
γ	1.5	5.2	3.4	3.4
η	2.8	15	0.63	0.63

The resonance width for low-frequency modes $\eta_{\mathbf{q}}$ averaged over wave vectors is given in the Debye approximation as follows:[143]

$$\eta = \frac{1}{N_0}\sum_{\mathbf{q}} \eta_{\mathbf{q}} = \frac{m_l \omega_l^4}{4\pi\rho c_t^3}, \tag{4.3.50}$$

where m_l is the reduced mass of low-frequency vibrations for adsorbed molecules, and ρ and c_t are density and the transverse sound velocity for the substrate material (see Eq. (4.2.25) and the alternative derivation of this formula in Appendix 1). Thus, Eq. (4.3.50) permits the parameter η to be independently estimated. As an example, for the system CO/NaCl(100) ($m_l \approx 2.68 \cdot 10^{-23}$ g, $\omega_l \approx 37.5$ cm^{-1}, $\rho \approx 2.18$ g/cm^3, $c_t \approx 2.2 \cdot 10^5$ cm/s), it is found that $\eta \approx 1.2$ cm^{-1}. With reference to this value, Eq. (4.3.47) provides a more adequate value of the parameter η than Eq. (4.3.49).

Consider vibrational excitations giving rise to small angular azimuthal deviations $\varphi_{\mathbf{q}}$ ($\mathbf{q} = \mathbf{k}a$, $\mathbf{k}$ is the two-dimensional wave vector of a collectivized excitation) of adsorbed molecules relative to the ground state AB (see Fig. 2.17) in the system CO/NaCl(100).[52] Expanding Eq. (2.4.2) in small φ_{nm} accurate to φ_{nm}^2 and adding the kinetic energy of orientational vibrations, we obtain, on subtracting the contribution of the ground state (2.4.7) and switching to normal coordinates $\varphi_{\mathbf{q}}$, the following expression for the excitation Hamiltonian in the harmonic approximation:

$$H_{ex} = \frac{1}{2}\sum_{\mathbf{q}} \left[I|\dot{\varphi}_{\mathbf{q}}|^2 + J(\mathbf{q})|\varphi_{\mathbf{q}}|^2 \right], \tag{4.3.51}$$

where $I = m(2d)^2\sin^2\theta$ is the molecular moment of inertia ($2d$ is the length of molecular bond) and the function

$$\begin{aligned} J(\mathbf{q}) = J(q_x, q_y) = \left[4B_3\cos^2\theta + (2B_3 + B_4)\sin^2\theta\right]\sin^2\theta\cos q_x \\ + 2(2B_3 + B_4)\left(\cos^2\theta - \sin^2\theta\right)\cos q_y \\ - 2B_4\cos^2\theta\sin^2\theta - 4(B_2 + 2B_3 + B_4 + 2B_5)\sin^4\theta - 8\Delta U_4 \end{aligned} \tag{4.3.52}$$

represents the dispersion law for orientational vibration provided that

$$J(\pi, 0) = -4(B_2 + 4B_3 + 2B_4 + 2B_5)\sin^4\theta - 8\Delta U_4 \geq 0. \tag{4.3.53}$$

At $\Delta U_4 = 0$ and $\sin\theta << 1$, the value $J(\pi,0)$ goes to zero, which gives rise to a Goldstone mode in a dipole-like system with the degenerate ground state, with the degeneration removed by thermodynamic fluctuations at nonzero temperatures.[70] At sufficiently small negative values of the reorientation barrier ΔU_4, an energy gap arises which corresponds to orientational vibrations of the frequency ω_l for isolated molecules, so that the dispersion law for orientational vibrations in the case of pure quadrupole-quadrupole interactions takes on the form:[52]

$$\omega_{\mathbf{q}}^2 = \omega_l^2 + 3\omega_Q^2\left[10 - 29t + 2(1-5t)\cos q_x - 8(1-2t)\cos q_y\right], \tag{4.3.54}$$

where $t = \sin^2\theta$ and

$$\omega_Q = \frac{1}{2d}\sqrt{\frac{U}{m}} \tag{4.3.55}$$

is the characteristic frequency for the quadrupole system. With the parameters ω_l = 37.5 cm^{-1}, ω_Q = 9.3 cm^{-1}, t = 0.1786 (θ= 25^0) inherent in the system CO/NaCl(100), the dependence $\omega_{\mathbf{q}}$ is plotted in Fig. 4.9.

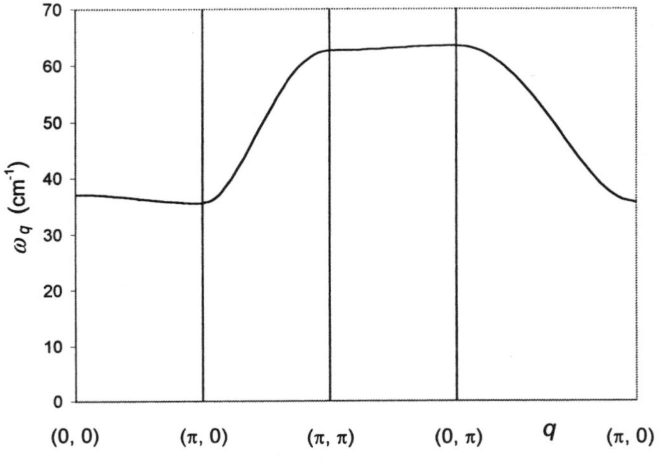

Fig. 4.9. Dispersion laws for orientational vibrations of inclined quadrupoles (θ= 25^0) on a square lattice relative to the ground state for the 2×1 phase of CO/NaCl(100).

Gently sloping spectral regions arise from small values of the coefficient of $\cos q_x$ and evidence for a stronger anisotropy of quadrupole interactions compared to dipole interactions. The average band halfwidth for collectivized orientational vibrations proves to be of the order 10 cm^{-1}. Thus, the orders of magnitudes of effective band halfwidths for low-frequency vibrations, $\Delta\omega_j$, (see Table 4.3) are consistent with the estimated band halfwidth for orientational vibrations of CO molecules which value is dictated by intermolecular quadrupole-quadrupole interactions. The values $\Delta\omega_S$ exceeding $\Delta\omega_A$ evidence for a lower density of low-frequency states in the band of symmetric vibrations compared to that of antisymmetric vibrations.

Appendix 1

Local and resonance states for a system of bound harmonic oscillators

The Hamiltonian function for a system of bound harmonic oscillators is, in the most general form, a sum of two positively definite quadratic forms composed of the particle momentum vectors and the Cartesian projections of particle displacements about equilibrium positions:

$$H = \sum_i \frac{\mathbf{p}_i^2}{2m_i} + \frac{1}{2}\sum_{ij,\alpha\beta} \Phi_{ij}^{\alpha\beta} u_i^\alpha u_j^\beta \, , \Phi_{ij}^{\alpha\beta} = \Phi_{ji}^{\beta\alpha} \, , \tag{A1.1}$$

where m_i is the mass of the ith particle; Φ denotes the force constants; α, β are the projections of corresponding vectors on the Cartesian coordinate axes (summation over Greek indices will be indicated explicitly because it appears in the expressions more than twice). The equations of motion for the displacements u_i^α that correspond to the Hamiltonian function (A1.1) are

$$m_i \ddot{u}_i^\alpha + \sum_{i\beta} \Phi_{ij}^{\alpha\beta} u_j^\beta = 0 \, . \tag{A1.2}$$

Both quadratic forms in Eq. (A1.1) can be diagonalized simultaneously by changing to new (normal) coordinates x:

$$u_i^\alpha = m_i^{-1/2} \sum_{q\nu} C_{iq}^{\alpha\nu} x_q^\nu = 0 \, , \tag{A1.3}$$

where $C_{iq}^{\alpha\nu}$ is the real orthogonal matrix with the indices i and q ranging from 1 to N (N is the number of particles in the system) and the indices α and ν ranging from 1 to d (d is the dimensionality of the space of displacements):

$$\sum_{i\alpha} C_{iq}^{\alpha\nu} C_{iq'}^{\alpha\nu'} = \delta_{qq'} \delta_{\nu\nu'} \, , \quad \sum_{q\nu} C_{iq}^{\alpha\nu} C_{jq}^{\beta\nu} = \delta_{ij} \delta_{\alpha\beta} \, . \tag{A1.4}$$

Since $p_i^\alpha = m_i \dot{u}_i^\alpha$, the kinetic energy in (A1.1) takes the form

$$\frac{1}{2}\sum_{i\alpha}\frac{\left(p_i^\alpha\right)^2}{m_i} = \sum_{\alpha q' \nu\nu'} C_{iq'}^{\alpha\nu} C_{iq'}^{\alpha\nu'} x_q^\nu x_{q'}^{\nu'} = \frac{1}{2}\sum_{q\nu}\left(\dot{x}_q^\nu\right)^2 . \tag{A1.5}$$

Using the formula (A1.3) we derive the potential energy as follows:

$$\frac{1}{2}\sum_{ij,\alpha\beta}\Phi_{ij}^{\alpha\beta} u_i^\alpha u_j^\beta = \frac{1}{2}\sum_{qq'\nu\nu'} x_q^\nu x_{q'}^{\nu'} \sum_{ij\alpha\beta}\left(m_i m_j\right)^{-1/2}\Phi_{ij}^{\alpha\beta} C_{iq}^{\alpha\nu} C_{jq'}^{\beta\nu'} . \tag{A1.6}$$

For the potential energy to be diagonal with respect to x_q^ν, we must require that

$$\sum_{ij\alpha\beta}\left(m_i m_j\right)^{-1/2}\Phi_{ij}^{\alpha\beta} C_{iq}^{\alpha\nu} C_{jq'}^{\beta\nu'} = \omega_{q\nu}^2 \delta_{qq'}\delta_{\nu\nu'} . \tag{A1.7}$$

By multiplying both sides of the Eq. (A1.7) by $C_{lq'}^{\gamma\nu'}$ and summing over q',ν' (with the Eq. (A1.4) taken into account), we have:

$$\sum_{i\alpha}\left(m_i m_l\right)^{-1/2}\Phi_{il}^{\alpha\gamma} C_{iq}^{\alpha\nu} = \omega_{q\nu}^2 C_{lq}^{\gamma\nu} , \tag{A1.8}$$

i.e., the columns number $q\nu$ of the matrix $C_{iq}^{\alpha\nu}$ are the eigenvectors of the matrix $\left(m_i m_l\right)^{-1/2}\Phi_{il}^{\alpha\gamma}$ with the eigenvalues $\omega_{q\nu}^2$ satisfying the equation

$$\det[\left(m_i m_l\right)^{-1/2}\Phi_{il}^{\alpha\gamma} - \omega^2\delta_{il}\delta_{\alpha\gamma}] = 0 . \tag{A1.9}$$

By substituting the equality (A1.7) into the relation (A1.6) and adding the result to Eq. (A1.5), we deduce the Hamiltonian function expressed in terms of normal coordinates:

$$H = \frac{1}{2}\sum_{q\nu}\left[\left(\dot{x}_q^\nu\right)^2 + \omega_{q\nu}^2\left(x_q^\nu\right)^2\right] \tag{A1.10}$$

and the corresponding equations of motion for x_q^ν :

$$\ddot{x}_q^\nu + \omega_{q\nu}^2 x_q^\nu = 0 \tag{A1.11}$$

which coincide with the equations of vibrations of a harmonic oscillator with the natural frequency $\omega_{q\nu}$.

An important and convenient characteristic of the system is its Green function (GF) that describes the response of the system to an instantaneous perturbation. We introduce the GF corresponding to Eq. (A1.11):

$$\ddot{G}_q^\nu(t) + \omega_{q\nu}^2 G_q^\nu(t) = -\delta(t), \tag{A1.12}$$

with the minus sign for the $\delta(t)$ corresponding to a standard definition by which (as will be shown below) $\mathrm{Im}G(\omega) < 0$. The solution of Eq. (A1.12) is:

$$G_q^\nu(t) = -\theta(t)\sin \omega_{q\nu} t / \omega_{q\nu} \,. \tag{A1.13}$$

The frequency Fourier component of the GF is defined by the relation:

$$G_q^\nu(\omega) = \int_{-\infty}^{\infty} G_q^\nu(t) e^{i\omega t} dt \,. \tag{A1.14}$$

If expression (A1.13) is directly substituted into Eq. (A1.14), the primitive of the integrand is easy to find, but the substitution of the integration limits, 0 and ∞, by the Newton-Leibnitz formula results in the uncertainty at the upper limit such as $\lim_{t\to\infty} e^{i(\omega \pm \omega_{q\nu})t}$. With the attenuation $\lambda \to +0$, the uncertainty is removed, and we arrive at:

$$G_q^\nu(\omega) = \left(\omega^2 - \omega_{q\nu}^2 + i0\,\mathrm{sign}\,\omega\right)^{-1}, \tag{A1.15}$$

where

$$\mathrm{sign}\,\omega = \begin{cases} 1 & \text{at } \omega > 0 \\ -1 & \text{at } \omega < 0 \end{cases} \tag{A1.16}$$

The presence of an infinitesimal, purely imaginary addition in the GF denominator turns out to be very important and to have a deep physical meaning. Its origin is related to the theta-function in Eq. (A1.13) that represents the causality principle: The reaction of a system can be caused only by perturbations at preceding instants.

Note that the relation (A1.15) (though without the imaginary addition to be completely defined later) is derivable by the Fourier transform of Eq. (A1.12). Since

$$G_q^{\nu}(t)=\frac{1}{2\pi}\int_{-\infty}^{\infty}G_q^{\nu}(\omega)e^{-i\omega t}d\omega,$$
$$\delta(t)=\frac{1}{2\pi}\int_{-\infty}^{\infty}e^{-i\omega t}d\omega \tag{A1.17}$$

(the last identity is one of the integral representations of the δ-function), then $G_q^{\nu}(\omega)=\left(\omega^2-\omega_{q\nu}^2\right)^{-1}$. We show how the inverse Fourier transform of Eq. (A1.15) performed by the theory of complex variable function takes us back to the original expression (A1.13). Since $\exp(-i\omega t)=\exp(-it\,\mathrm{Re}\,\omega\, t)\exp(t\,\mathrm{Im}\,\omega\, t)$, the integrals over the semicircle of an infinitely large radius are zeroes in the upper and the lower half-planes at respectively $t < 0$ and $t > 0$. Therefore, the integral along the real axis ω may be replaced by the integrals round the following closed circuits (see Fig. A1.1):

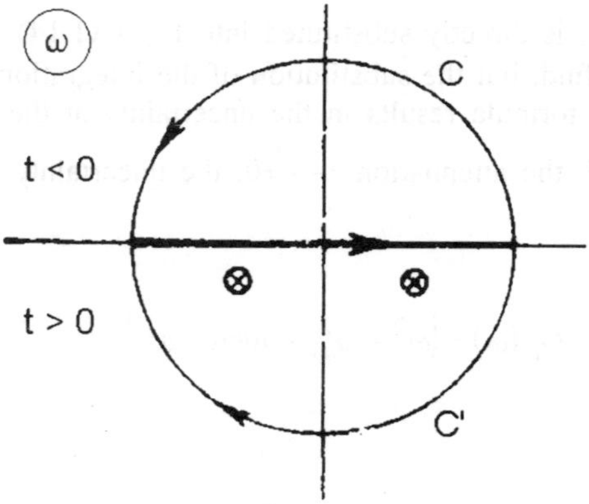

Fig. A1.1. Integration contours in formula (A1.18).

$$G_q^\nu(t) = \begin{cases} \dfrac{1}{2\pi} \displaystyle\int\limits_C G_q^\nu(\omega) e^{-i\omega t} d\omega, & t < 0, \\ \dfrac{1}{2\pi} \displaystyle\int\limits_{C'} G_q^\nu(\omega) e^{-i\omega t} d\omega, & t > 0. \end{cases} \tag{A1.18}$$

The poles of the integrand $\omega = \pm\omega_{q\nu} - i0$ lie in the lower half-plane (see Fig. A1.1), so that at t < 0 the case is reduced to an integral of the analytic function round a closed circuit in the upper half-plane which is equal to zero. At t > 0, the value of the integral is found by the theory of residues:

$$G_q^\nu(t) = -2\pi i \mathrm{Res}\left[\frac{1}{2\pi} G_q^\nu(\omega) e^{-i\omega t}\right] = -\frac{1}{\omega_{q\nu}} \sin \omega_{q\nu} t\,, \tag{A1.19}$$

so that we again arrive at (A1.13). Thus, the infinitesimal imaginary addition at the pole of the GF (A1.15) really arises from the theta-function in Eq. (A1.13).

We now introduce the GF $G_{ij}^{\alpha\beta}(t)$ for the displacements u_i^α that satisfies the equation

$$m_i \ddot{G}_{ij}^{\alpha\beta}(t) + \sum_{j'\gamma} \Phi_{ij'}^{\alpha\gamma} G_{j'j}^{\gamma\beta}(t) = -\delta(t)\delta_{ij}\delta_{\alpha\beta} \tag{A1.20}$$

and determines the response of the displacements $u_i^\alpha(t)$ to the external forces F_i^α (which should be substituted in the right-hand side of Eq. (A1.2)):

$$u_i^\alpha(t) = -\sum_{j\beta} \int\limits_{-\infty}^{\infty} G_{ij}^{\alpha\beta}(t-t') F_j^\beta(t') dt'\,. \tag{A1.21}$$

It is straightforward to show that $G_{ij}^{\alpha\beta}(t)$ is related to the GF $G_q^\nu(t)$ by the relation

$$G_{ij}^{\alpha\beta}(t) = (m_i m_j)^{-1/2} \sum_{q\nu} G_q^\nu(t) C_{iq}^{\alpha\nu} C_{jq}^{\beta\nu}\,. \tag{A1.22}$$

For this purpose, we only need to substitute Eq. (A1.22) into (A1.20) so as to deduce, using formulae (A1.4) and (A1.8), Eq. (A1.12) for $G_q^\nu(t)$. Formula (A1.22) relates the GF of the non-diagonal representation for the initial displacements u_i^α to the GF of the diagonal representation for normal coordinates x_q^ν. Evidently, the same relation holds for the frequency Fourier components of the GF, so that

$$G_{ij}^{\alpha\beta}(\omega)=\left(m_i m_j\right)^{-1/2}\sum_{q\nu}\frac{C_{iq}^{\alpha\nu}C_{jq}^{\beta\nu}}{\omega^2-\omega_{q\nu}^2+i0\,\mathrm{sign}\,\omega}. \tag{A1.23}$$

Invoking Sokhotskii formula, we obtain the imaginary part of the GF:

$$\mathrm{Im}\,G_{ij}^{\alpha\beta}(\omega)=-\pi\left(m_i m_j\right)^{-1/2}\mathrm{sign}\,\omega\sum_{q\nu}C_{iq}^{\alpha\nu}C_{jq}^{\beta\nu}\delta\left(\omega^2-\omega_{q\nu}^2\right). \tag{A1.24}$$

Two significant properties of $\mathrm{Im}\,G_{ij}^{\alpha\beta}(\omega)$ following from Eq. (A1.24) deserve mention:

$$\int_{-\infty}^{\infty}\omega\,\mathrm{Im}\,G_{ij}^{\alpha\beta}(\omega)\,d\omega=$$

$$-\pi\left(m_i m_j\right)^{-1/2}\mathrm{sign}\,\omega\sum_{q\nu}C_{iq}^{\alpha\nu}C_{jq}^{\beta\nu}\int_{-\infty}^{\infty}|\omega|\delta\left(\omega^2-\omega_{q\nu}^2\right)d\omega=-\frac{\pi}{m_i}\delta_{ij}\delta_{\alpha\beta}, \tag{A1.25}$$

$$\sum_{i\alpha}m_i\,\mathrm{Im}\,G_{ij}^{\alpha\beta}(\omega)=-\pi\,\mathrm{sign}\,\omega\sum_{q\nu}\delta\left(\omega^2-\omega_{q\nu}^2\right). \tag{A1.26}$$

Introduce a normalized distribution function for squared frequencies alternatively called a state density:

$$\rho\left(\omega^2\right)=\frac{1}{Nd}\sum_{q\nu}\delta\left(\omega^2-\omega_{q\nu}^2\right),\quad \int_0^{\infty}\rho\left(\omega^2\right)d\omega^2=1 \tag{A1.27}$$

(N is the total number of particles; d is the dimensionality of the space of displacements). According to Eq. (A1.26), this function is related to $\operatorname{Im} G_{ij}^{\alpha\beta}(\omega)$ at $\omega > 0$:

$$\rho\left(\omega^2\right) = -\frac{1}{\pi N d} \sum_{i\alpha} m_i \operatorname{Im} G_{ii}^{\alpha\alpha}(\omega). \tag{A1.28}$$

The physical meaning of the function $\rho(\omega^2)$ is that the quantity $\rho(\omega^2)d\omega^2$ specifies the number of squared frequencies $\omega_{q\nu}^2$ (divided by their total number Nd) falling within the interval from ω^2 to $\omega^2 + \delta\omega^2$. This essential characteristic of the frequency spectrum is also defined by the GF of the system.

The above treatment needs to be summarized. Transform (A1.3) for displacements and relation (A1.22) for the GF make it possible to express the quantities desired in terms of simple solutions of Eq. (A1.11) for independent vibrations of harmonic oscillators and their GFs (see Eqs. (A1.13) and (A1.15)). In so doing, the transformation matrices $C_{iq}^{\alpha\nu}$ and the frequencies $\omega_{q\nu}$ are found to be the solutions of the eigenvalue problem (see Eq. (A1.8)). To explicitly derive these quantities is impossible in the general case, as it would require solving a system of Nd linear equations or a single equation in ω^2 (see Eq. (A1.9)) of the degree Nd. Below we will examine three particular cases of importance in which the natural frequency spectrum is deduced in an explicit form.

Consider a system of two bound one-dimensional oscillators whose interaction with each other and with the wall is effected by two different springs (see Fig. A1.2).

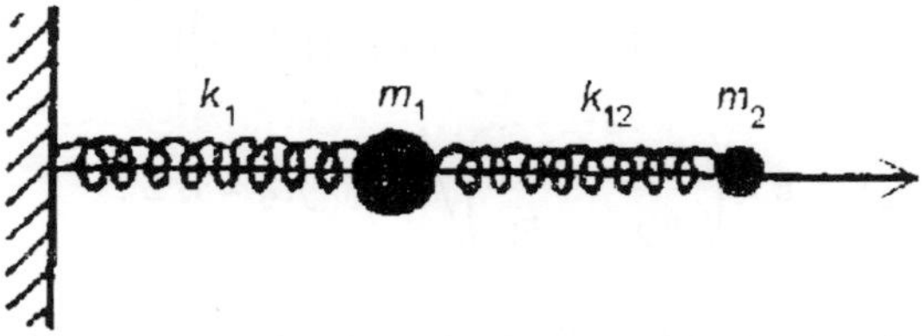

Fig. A1.2. Two coupled oscillators.

Then in the initial Hamiltonian function (A1.1), all the indices, α and β, of the projections onto the Cartesian axes are the same and thus can be omitted, while the indices of particle numbers assume only two values, $i = 1, 2$. Designating the spring force constants as k_1 and k_{12}, we come to

$$H = \frac{p_1^2}{2m_1} + \frac{p_2^2}{2m_2} + \frac{1}{2}k_1 u_1^2 + \frac{1}{2}k_{12}(u_2 - u_1)^2 , \tag{A1.29}$$

so that the force constants Φ_{ij} in Eq. (A1.1) are:

$$\Phi_{11} = k_1 + k_{12}, \quad \Phi_{12} = -k_{12}, \quad \Phi_{22} = k_{12} . \tag{A1.30}$$

The equation (A1.8) becomes as follows:

$$\left(\frac{\Phi_{11}}{m_1} - \omega_q^2\right) C_{1q} + \frac{\Phi_{12}}{\sqrt{m_1 m_2}} C_{2q} = 0 , \tag{A1.31}$$

$$\frac{\Phi_{12}}{\sqrt{m_1 m_2}} C_{1q} + \left(\frac{\Phi_{22}}{m_2} - \omega_q^2\right) C_{2q} = 0 . \tag{A1.32}$$

Hence, we have:

$$\omega_q^2 = \frac{1}{2}\left[\frac{\Phi_{11}}{m_1} + \frac{\Phi_{22}}{m_2} + (-1)^q D\right], \quad D \equiv \left[\left(\frac{\Phi_{11}}{m_1} - \frac{\Phi_{22}}{m_2}\right)^2 + 4\frac{\Phi_{12}^2}{m_1 m_2}\right]^{1/2} , \tag{A1.33}$$

$$\begin{aligned} C_{jq} &= (-1)^{j(q-1)}\left[\frac{1}{2}\left(1 + (-1)^{j+q} B\right)\right]^{1/2} , \\ B &\equiv \left(\frac{\Phi_{22}}{m_2} - \frac{\Phi_{11}}{m_1}\right) \Big/ D, \qquad j, q = 1, 2 . \end{aligned} \tag{A1.34}$$

We simplify the expression (A1.33) for a case when $k_1 \sim k_{12}$ and $m_2 << m_1$. It is easily expanded in the small parameter

$$\frac{m}{M} = \frac{m_1 m_2}{(m_1 + m_2)^2} \approx \frac{m_2}{m_1} - 2\left(\frac{m_2}{m_1}\right)^2 , \quad m_2 << m_1 \tag{A1.35}$$

which is the ratio of the reduced mass $m = m_1m_2/M$ of two particles to their total mass $M = m_1 + m_2$. As a result, we obtain, accurate to terms of the order $(m/M)^2$:

$$m_1 \approx M\left(1-\frac{m}{M}-\frac{m^2}{M^2}\right), \quad m_2 \approx m\left(1+\frac{m}{M}+2\frac{m^2}{M^2}\right) \tag{A1.36}$$

and the formula (A1.33) takes the form:

$$\omega_1^2 \approx \frac{k_1}{M}\left(1-\frac{m^2}{M^2}\frac{k_1}{k_{12}}\right), \quad \omega_2^2 \approx \frac{k_{12}}{m}\left(1+\frac{m^2}{M^2}\frac{k_1}{k_{12}}\right). \tag{A1.37}$$

So, at $m_2 << m_1$, the natural frequencies of the system correspond to the independent vibrations of the mass M on the spring k_1 and of the reduced mass m on the spring k_{12}. As the parameter B tends to unity for $m_2 << m_1$, the relative displacement of particles 1 and 2 is approximately described by the normal coordinate x_2: $u_2 - u_1 \approx x_2/\sqrt{m}$.

Thus, in view of the inequality $m_2 << m_1$, we derive a physically obvious result: the motion of a light particle 2 can approximately be regarded as vibrations about a relatively immobile particle 1. The squared frequency of these vibrations would be equal to k_{12}/m_2 thus differing (by virtue of the approximate equality (A1.36)) from k_{12}/m_2 by the terms of the order m/M. Besides, the second formula in (A1.37) shows that the approximate notation of the sought-for frequency as k_{12}/m is preferable because it is valid accurate to terms of the order $(m/M)^2$.

As another example illustrating an explicit switch to normal coordinates, we consider a three-dimensional monoatomic simple lattice. In such a system, masses of all particles are the same and the positions of their stable equilibria are at the lattice sites which are given by radius vectors **n** (called lattice vectors). Instead of an unsystematic particle numbering ($i = 1, 2, ..., N$), it is now convenient to distinguish them by the lattice sites they belong to and to designate them by the index **n**. The periodicity of the particle positions implies that the force constant $\Phi^{\alpha\beta}_{\mathbf{nn}'}$ for a pair of particles **n** and **n'** should be the same as that for a pair of particles **n+a** and **n'+a**, where **a** is an arbitrary lattice vector. Thus, $\Phi^{\alpha\beta}_{\mathbf{nn}'} = \Phi^{\alpha\beta}_{\mathbf{n+a,n'+a}}$ but, by the arbitrariness of **a**, we can put **a** = -**n'** and then $\Phi^{\alpha\beta}_{\mathbf{nn}'} = \Phi^{\alpha\beta}_{\mathbf{n-n',0}} \equiv \Phi(\mathbf{n}-\mathbf{n}')$, i.e., the force constants are dependent only on the difference **n-n'**. The displacement of the lattice as a whole should not result in changed force constants, which is equivalent to the condition:

$$\sum_{\mathbf{n}'} \Phi^{\alpha\beta}(\mathbf{n}-\mathbf{n}') = 0 . \tag{A1.38}$$

Note that in the above-discussed model (see Fig. A1.2), the particle 1 is "bound" by the spring k_1 to an immobile wall, so that the condition (A1.38) was not met (see Eq. (A1.30)).

The Hamiltonian function

$$H = \frac{1}{2m}\sum_{\mathbf{n}} \mathbf{p}_{\mathbf{n}}^2 + \frac{1}{2}\sum_{\mathbf{nn}'} \Phi^{\alpha\beta}(\mathbf{n}-\mathbf{n}') u_{\mathbf{n}}^{\alpha} u_{\mathbf{n}'}^{\beta} = 0 \tag{A1.39}$$

is easily diagonalized by the transform (A1.3) in which $m_i = m$, $\mathbf{q}$ is the vector running through a quasi-continuous spectrum of values (at $N >> 1$) provided cyclic boundary conditions:

$$\mathbf{u}_{\mathbf{n}+N_i\mathbf{a}_i} = \mathbf{u}_{\mathbf{n}}, \quad i = 1, 2, 3 \tag{A1.40}$$

($\mathbf{a}_1$, $\mathbf{a}_2$, $\mathbf{a}_3$ are the basis lattice vectors; N_1, N_2, N_3 are the large integers that fix the main area for a lattice of $N = N_1N_2N_3$ sites). The matrix $C_{\mathbf{nq}}^{\alpha\nu}$ has the complex matrix elements

$$C_{\mathbf{nq}}^{\alpha\nu} = \frac{1}{\sqrt{N}} e_{\alpha\nu}(\mathbf{q}) \exp(i\mathbf{q}\cdot\mathbf{n}) \tag{A1.41}$$

obeying the unitarity relations

$$\sum_{\mathbf{n}\alpha} C_{\mathbf{nq}}^{\alpha\nu} C^{*\alpha\nu'}_{\mathbf{nq}'} = \delta_{\mathbf{qq}'}\delta_{\nu\nu'}, \quad \sum_{\mathbf{q}\nu} C_{\mathbf{nq}}^{\alpha\nu} C^{*\beta\nu}_{\mathbf{n}'\mathbf{q}} = \delta_{\mathbf{nn}'}\delta_{\alpha\beta} \tag{A1.42}$$

which generalize the orthogonality conditions (A1.4) for real matrices. As a result, the transform (A1.3) becomes:

$$u_{\mathbf{n}}^{\alpha} = \frac{1}{\sqrt{mN}} \sum_{\mathbf{q}\nu} x_{\mathbf{q}}^{\nu} e_{\alpha\nu}(\mathbf{q}) \exp(i\mathbf{q}\cdot\mathbf{n}) \tag{A1.43}$$

and then formula (A1.40) is equivalent to the condition:

$$\mathbf{q}\cdot\mathbf{a} = \frac{2\pi}{N_i} g_i, \quad g_i = 0, \pm 1, \pm 2, \ldots \tag{A1.44}$$

The solution of this equation for $\mathbf{q}$ is well known:

$$\mathbf{q} = \sum_{i=1}^{3} \frac{g_i}{N_i}\mathbf{b}_i, \ \mathbf{b}_1 = \frac{2\pi}{V_0}\mathbf{a}_2 \times \mathbf{a}_3, \quad \mathbf{b}_2 = \frac{2\pi}{V_0}\mathbf{a}_3 \times \mathbf{a}_1,$$
$$\mathbf{b}_3 = \frac{2\pi}{V_0}\mathbf{a}_1 \times \mathbf{a}_2, \quad V_0 = \mathbf{a}_1\mathbf{a}_2\mathbf{a}_3 = \mathbf{a}_1 \cdot [\mathbf{a}_2 \times \mathbf{a}_3] \tag{A1.45}$$

and explicitly defines the spectrum of values for $\mathbf{q}$. The vectors $\mathbf{b}_i$ are called the basis vectors of a reciprocal lattice and satisfy (as it follows from Eq. (A1.45)) the relations:

$$\mathbf{a}_i \cdot \mathbf{b}_i = 2\pi\delta_{ij}\,. \tag{A1.46}$$

The quantity V_0 is equal to the volume of a lattice unit cell. Since $\exp(i\mathbf{q}\cdot\mathbf{n})$ is a periodic function with the period 2π, the integer values g_i may be restricted by the inequalities:

$$-N_i/2 < g_i \le N_i/2, \quad i = 1, 2, 3\,. \tag{A1.47}$$

The range of $\mathbf{q}$ corresponding to the above inequalities is referred to as the first Brillouin zone. It is now easy to prove the following two equalities of significance:

$$\frac{1}{N}\sum_{\mathbf{n}} \exp[i(\mathbf{q}-\mathbf{q}')\cdot\mathbf{n}] = \delta_{\mathbf{qq}'}, \quad \frac{1}{N}\sum_{\mathbf{q}} \exp[i(\mathbf{n}-\mathbf{n}')\cdot\mathbf{q}] = \delta_{\mathbf{nn}'} \tag{A1.48}$$

in which the summation variables $\mathbf{n}$ and $\mathbf{q}$ run through exactly N above-specified values. Substituting Eq. (A1.41) into Eq. (A1.42) with regard to Eq. (A1.48), we obtain the unitarity relations for the matrices $e_{\alpha\nu}(\mathbf{q})$:

$$\sum_{\alpha} e_{\alpha\nu}(\mathbf{q}) e^*_{\alpha\nu'}(\mathbf{q}) = \delta_{\nu\nu'}, \quad \sum_{\nu} e_{\alpha\nu}(\mathbf{q}) e^*_{\beta\nu}(\mathbf{q}) = \delta_{\alpha\beta}\,. \tag{A1.49}$$

Eq. (A1.8) that defines the frequency spectrum $\omega_\nu(\mathbf{q})$ for the system assumes the form:

$$\frac{1}{m}\sum_{\beta}\Phi^{\alpha\beta}(\mathbf{q})e_{\beta\nu}(\mathbf{q})=\omega_\nu^2(\mathbf{q})e_{\alpha\nu}(\mathbf{q}), \tag{A1.50}$$

where the third-rank matrices

$$\Phi^{\alpha\beta}(\mathbf{q})=\sum_{\mathbf{n}}\Phi^{\alpha\beta}(\mathbf{n})e^{-i\mathbf{q}\cdot\mathbf{n}} \tag{A1.51}$$

are the coefficients for the expansion of force constants in a Fourier series in terms of $\mathbf{q}$:

$$\Phi^{\alpha\beta}(\mathbf{n})=\frac{1}{N}\sum_{\mathbf{q}}\Phi^{\alpha\beta}(\mathbf{q})e^{i\mathbf{q}\cdot\mathbf{n}}. \tag{A1.52}$$

Note that $\Phi^{\alpha\beta}(\mathbf{n})=\Phi^{\beta\alpha}(-\mathbf{n})$ due to the symmetry of the matrices $\Phi_{ij}^{\alpha\beta}$ (see Eq. (A1.1)). Further, $\Phi^{\alpha\beta}(\mathbf{q})=\Phi^{*\beta\alpha}(-\mathbf{q})$ since values $\Phi^{\alpha\beta}(\mathbf{n})$ are real. As a consequence,

$$\Phi^{\alpha\beta}(\mathbf{q})=\Phi^{\beta\alpha}(-\mathbf{q}), \tag{A1.53}$$

so that the matrix $\Phi^{\alpha\beta}(\mathbf{q})$ is Hermitian, its eigenvectors $\mathbf{e}_\nu(\mathbf{q})$ are orthogonal (in the sense of the first equality in Eq. (A1.49)), and eigenvalues $\omega_\nu^2(\mathbf{q})$ are real (the positivity of $\omega_\nu^2(\mathbf{q})$ follows from the positive definiteness of the quadratic form for the potential energy). We subject Eq. (A1.50) to the complex conjugation operation and take into consideration the property (A1.53). Then the quantity $e_{\alpha\nu}^*(\mathbf{q})$ will represent the eigenvector $e_{\alpha\nu}(-\mathbf{q})$ with the same eigenvalue $\omega_\nu^2(\mathbf{q})$, i.e.,

$$\omega_\nu^2(-\mathbf{q})=\omega_\nu^2(\mathbf{q}), \qquad e_{\alpha\nu}(-\mathbf{q})=e_{\alpha\nu}^*(\mathbf{q}). \tag{A1.54}$$

For inversion-center lattices, a stronger condition, $\Phi^{\alpha\beta}(\mathbf{n}) = \Phi^{\alpha\beta}(-\mathbf{n})$, is valid, so that the matrix $\Phi^{\alpha\beta}(\mathbf{q})$ is real and symmetric and hence the eigenvectors $\mathbf{e}_\nu(\mathbf{q})$ may be considered to be real.

The GF (A1.23) of the system in question is

$$G^{\alpha\beta}(\mathbf{n},\omega) = \frac{1}{mN}\sum_{\mathbf{q}\nu}\frac{e_{\alpha\nu}(\mathbf{q})e^*_{\beta\nu}(\mathbf{q})}{\omega^2 - \omega_\nu^2(\mathbf{q}) + i0\operatorname{sign}\omega}\exp(i\mathbf{q}\cdot\mathbf{n}) \tag{A1.55}$$

and on expanding it in a Fourier series in $\mathbf{q}$ (as in Eqs. (A1.51) and (A1.52)), the resulting coefficients appear as

$$G^{\alpha\beta}(\mathbf{q},\omega) = \frac{1}{m}\sum_{\nu}\frac{e_{\alpha\nu}(\mathbf{q})e^*_{\beta\nu}(\mathbf{q})}{\omega^2 - \omega_\nu^2(\mathbf{q}) + i0\operatorname{sign}\omega}. \tag{A1.56}$$

The quantity (A1.56) at $\omega > 0$ is associated with the so-called spectral function

$$S(\mathbf{q},\omega^2) = -\frac{m}{\pi d}\sum_{\alpha}\operatorname{Im}G^{\alpha\alpha}(\mathbf{q},\omega) = \frac{1}{d}\sum_{\nu}\delta\left(\omega^2 - \omega_\nu^2(\mathbf{q})\right) \tag{A1.57}$$

which characterizes the absorption spectrum if the radiation with the frequency ω and the wave vector $\mathbf{q}$ is absorbed by the system concerned. The definition of $S(\mathbf{q},\omega^2)$ gives its relation to the state density $\rho(\omega^2)$ (see Eqs. (A1. 27) and (A1. 28)) and the normalization condition:

$$\rho(\omega^2) = \frac{1}{N}\sum_{\mathbf{q}} S(\mathbf{q},\omega^2), \qquad \int_0^\infty S(\mathbf{q},\omega^2)\,d\omega^2 = 1. \tag{A1.58}$$

Importantly, from Eq. (A1.51) it follows $\Phi^{\alpha\beta}(\mathbf{q}=0) = 0$ (in view of Eq. (A1.38)) and hence Eq. (2.50) yields $\omega_\nu^2(\mathbf{q}=0) = 0$. Vibrations characterized by this property are said to be acoustic. The symmetry properties expressed by Eq. (A1.53) result in the fact that there are no terms linear in $\mathbf{q}$ in the expansion of $\Phi^{\alpha\nu}(\mathbf{q})$ in small $\mathbf{q}$. The expansion will, therefore, begin with quadratic terms which have the following general form for an isotropic lattice:

$$\frac{1}{m}\Phi^{\alpha\beta}(\mathbf{q}) = \lambda_1^2 q_\alpha q_\beta + \lambda_2^2 q^2 \delta_{\alpha\beta}. \tag{A1.59}$$

Substituting Eq. (A1. 59) into Eq. (A1. 50), it is readily seen that

$$\omega_1^2(\mathbf{q}) = \left(\lambda_1^2 + \lambda_2^2\right) q^2, \quad \mathbf{e}_1(\mathbf{q}) = \mathbf{q}/q; \tag{A1.60}$$

$$\omega_{2,3}^2(\mathbf{q}) = \lambda_2^2 q^2, \quad \mathbf{e}_2(\mathbf{q}), \mathbf{e}_3(\mathbf{q}) \perp \mathbf{e}_1(\mathbf{q}). \tag{A1.61}$$

The time dependence of the normal coordinates $x_{\mathbf{q}}^{\nu}$ obeying Eq. (A1.11) is given by the factor $\exp(-i\omega_\nu(\mathbf{q})\,t)$, so that the displacements of the particle, $\mathbf{u}_\mathbf{n}$, corresponding to this normal vibration are related, according to Eq. (A1.43), to the coordinates $\mathbf{n}$ and to the time t by the cofactor $\exp[i(\mathbf{q}\cdot\mathbf{n} - \omega_\nu(\mathbf{q})\,t)]$. This cofactor describes a plane wave propagating in the direction of the vector $\mathbf{q}$. The phase velocity of the wave propagation is specified by $c_\nu = \omega_\nu(\mathbf{q})/q$. For small q, these waves are acoustic (because the wavelength $2\pi/q$ notably exceeds the size of the lattice unit cell) and the expansion coefficients λ_1 and λ_2 in Eq. (A1.59) are related to acoustic wave velocities. The solution (A1.60) accounts for longitudinal waves (in which the particle displacements are parallel to the wave propagation direction), and Eq. (A1.61) describes two degenerate (since $\omega_2 = \omega_3$) transverse waves. The velocities of the longitudinal (c_l) and the transverse (c_t) acoustic waves are expressible in terms of λ_1 and λ_2 taken from Eqs. (A1.60) and (A1.61):

$$c_l = \sqrt{\lambda_1^2 + \lambda_2^2}, \quad c_t = \lambda_2 \tag{A1.62}$$

with $c_l > c_t$. The dependence of the frequency on the wave vector $\mathbf{q}$ is called the dispersion law; it is linear and the corresponding distribution function for squared frequencies (see Eq. (A1.27)) can easily be calculated by replacing the integral over $\mathbf{q}$ (with $N \to \infty$) for the sum over $\mathbf{q}$. We demonstrate this technique below.

The summation over $\mathbf{q}$ implies summing thrice over the integers g_i ($i = 1, 2, 3$) in Eq. (A1.45). Changing to the integration over $dg_1 dg_2 dg_3$, note that the Jacobian of the transformation from g_i to q_α is (according to Eqs. (A1.44) and (A1.45)):

$$\frac{\partial(g_1, g_2, g_3)}{\partial(q_x, q_y, q_z)} = \frac{N_1 N_2 N_3}{(2\pi)^3} V_0 = \frac{V}{(2\pi)^3}, \tag{A1.63}$$

where $V = NV_0$ is the total volume of the main area. As a result, we are led to a frequently used relation:

$$\sum_{\mathbf{q}} \ldots = \frac{V}{(2\pi)^3} \int d\mathbf{q} \ \ldots \tag{A1.64}$$

(In the case of a d-dimensional space, a pre-integral cofactor changes to $\widetilde{V}/(2\pi)^d$, with $\widetilde{V}$ denoting the "volume" of the main area in a d-dimensional space). The quantity $\rho(\omega^2)$ is now easy to find by Eq. (A1.27) (or Eqs. (A1.57) and (A1.58)):

$$\begin{aligned}\rho(\omega^2) &= \frac{1}{3N}\sum_{\mathbf{q}\nu} \delta\left(\omega^2 - \omega_\nu^2(\mathbf{q})\right) = \frac{V_0}{3(2\pi)^3}\sum_{\nu} \int d\mathbf{q}\,\delta\left(\omega^2 - \omega_\nu^2(\mathbf{q})\right) \\ &= \frac{V_0}{6\pi^2} \int q^2 dq \left[\delta\left(\omega^2 - c_l^2 q^2\right) + 2\delta\left(\omega^2 - c_t^2 q^2\right)\right] = \frac{V_0}{4\pi^2 \bar{c}^3}\,\omega,\end{aligned} \tag{A1.65}$$

$$\frac{1}{\bar{c}^3} = \frac{1}{3}\left(\frac{1}{c_l^3} + \frac{1}{c_t^3}\right). \tag{A1.66}$$

Of course, as q increases, the dispersion law $\omega_\nu(\mathbf{q})$ deviates from linear, so that the integral over q in Eq. (A1.65) should be limited from above. It is reasonable to so limit the admissible frequencies ω that the state density normalization (A1.27) is preserved:

$$\int_0^{\omega_\mathrm{D}^2} \rho\left(\omega^2\right) d\omega^2 = \frac{V_0}{6\pi^2 \bar{c}^3}\,\omega_\mathrm{D}^3 = 1\,. \tag{A1.67}$$

Hence, we obtain:

$$\omega_\mathrm{D} = \bar{c} q_\mathrm{D}, \quad q_\mathrm{D} = \left(6\pi^2/V_0\right)^{1/3}. \tag{A1.68}$$

Such an approximate description of acoustic vibrations is referred to as the Debye approximation and the limiting frequency ω_D is called the Debye frequency. The corresponding limiting value of the wave vector $\mathbf{q}$ is of the same order as the reciprocal lattice constant.

The GF (A1.55) satisfies the equation:

$$m\omega^2 G^{\alpha\beta}_{\mathbf{nn}'}(\omega) - \sum_{\mathbf{n}''\gamma} \Phi^{\alpha\gamma}(\mathbf{n}-\mathbf{n}'') G^{\gamma\beta}_{\mathbf{n}''\mathbf{n}'}(\omega) = \delta_{\mathbf{nn}'}\delta_{\alpha\beta} \tag{A1.69}$$

which represents a frequency Fourier transform of Eq. (A1.20) for a periodic lattice of particles with identical masses.

Suppose that the particle at the site $\mathbf{n} = 0$ has the mass $m_C = m - \Delta m$ different from the mass m of the other particles. The equation determining GF $\widetilde{G}^{\alpha\beta}_{\mathbf{nn}'}(\omega)$ for such a system is conveniently written as follows:

$$m\omega^2 \widetilde{G}^{\alpha\beta}_{\mathbf{nn}'}(\omega) - \sum_{\mathbf{n}''\gamma} \Phi^{\alpha\gamma}(\mathbf{n}-\mathbf{n}'') \widetilde{G}^{\gamma\beta}_{\mathbf{n}''\mathbf{n}'}(\omega) = \Delta m\omega^2 \widetilde{G}^{\alpha\beta}_{\mathbf{nn}'}(\omega)\delta_{\mathbf{n}0} + \delta_{\mathbf{nn}'}\delta_{\alpha\beta}\,. \tag{A1.70}$$

The left-hand sides of Eqs. (A1.69) and (A1.70) are of the same form, which enables the perturbed GF with the mass defect Δm to be readily expressed in terms of the unperturbed one:

$$\widetilde{G}^{\alpha\beta}_{\mathbf{nn}'}(\omega) = G^{\alpha\beta}_{\mathbf{nn}'}(\omega) + \Delta m\omega^2 \sum_{\gamma} G^{\alpha\gamma}_{\mathbf{n}0}(\omega)\widetilde{G}^{\gamma\beta}_{0\mathbf{n}'}(\omega)\,. \tag{A1.71}$$

Assuming $\mathbf{n} = 0$ in Eq. (A1.71), we come to:

$$\sum_{\gamma} \left[\delta_{\alpha\gamma} - \Delta m\omega^2 G^{\alpha\gamma}_{\mathbf{n}0}(\omega)\right]\widetilde{G}^{\gamma\beta}_{0\mathbf{n}'}(\omega) = G^{\alpha\beta}_{0\mathbf{n}'}(\omega)\,. \tag{A1.72}$$

For an isotropic lattice, the matrix $G^{\alpha\gamma}_{00}(\omega)$ is also isotropic and, on the strength of Eqs. (A1.27) and (A1.55), expressible in terms of the state density:

$$G^{\alpha\gamma}_{00}(\omega) = G_{00}(\omega)\delta_{\alpha\gamma}, \quad G_{00}(\omega) = \frac{1}{3mN}\sum_{\mathbf{q}\nu}\left[\omega^2 - \omega^2_\nu(\mathbf{q}) + i0\ \mathrm{sign}\,\omega\right]^{-1}$$
$$= \frac{1}{m}\int_0^\infty \frac{\rho\left(\widetilde{\omega}^2\right)d\widetilde{\omega}^2}{\omega^2 - \widetilde{\omega}^2 + i0\ \mathrm{sign}\,\omega} = m^{-1}\left[P\left(\omega^2\right) - i\pi\rho\left(\omega^2\right)\right]. \tag{A1.73}$$

In the latter equality, the Sokhotskii formula is used and the designation $P(\omega^2)$ is introduced for the principal integral value:

$$P\left(\omega^2\right) \equiv \int_0^{\infty} \frac{\rho\left(\tilde{\omega}^2\right) d\tilde{\omega}^2}{\omega^2 - \tilde{\omega}^2} . \tag{A1.74}$$

Thus, the lattice isotropy permits a straightforward relation between the perturbed and the unperturbed GF which is obtained without solving the system of linear equations (A1.72) in the general case:

$$\tilde{G}_{0\mathbf{n}}^{\alpha\beta}(\omega) = G_{0\mathbf{n}}^{\alpha\beta}(\omega) / [1 - \Delta m \omega^2 G_{00}(\omega)] . \tag{A1.75}$$

By the very definition of the GF, the real parts of the poles of its frequency Fourier component correspond to natural frequencies of the system (see, for example, Eqs. (A1.23) or (A1.55)). Consequently, the spectrum of natural frequencies of the perturbed system, ω_p, should fit the equation

$$\frac{\omega_p^2}{3N} \sum_{\mathbf{q}\nu} \left[\omega_p^2 - \omega_\nu^2(\mathbf{q})\right]^{-1} = \frac{m}{\Delta m} \tag{A1.76}$$

corresponding to the vanishing denominator in the GF (A1.75), with $\mathrm{Re}G_{00}(\omega)$ taken from Eq. (A1.73). The qualitative behavior for solutions of Eq. (A1.76) is conveniently exemplified by several equidistantly disposed nondegenerate frequencies $\omega_\nu^2(\mathbf{q})$. A graphic solution of this kind, with $3N = 6$, is presented in Fig A1.3.

First of all, it is noteworthy that the parameter $m/\Delta m = (1 - m_C/m)^{-1}$ can assume any negative value at $m_C > m$ and any positive value greater than unity at $m_C < m$. In this physically admissible range of the values $m/\Delta m$, Eq. (A1.76) has exactly $3N$ roots, as is to be expected for a system with $3N$ vibrational degrees of freedom. At $m_C \to m$, we obtain $m/\Delta m \to \pm\infty$ and the spectrum of natural frequencies coincides with $\omega_\nu(\mathbf{q})$. At $m_C \neq m$, each of the values ω_p^2 is necessarily between two neighboring squared frequencies $\omega_\nu^2(\mathbf{q})$ of the unperturbed system, i.e., the spectra ω_p^2 and $\omega_\nu^2(\mathbf{q})$ alternate. If s quantities $\omega_\nu^2(\mathbf{q})$ (with different **q** and ν) have the same value (the value of $\omega_\nu^2(\mathbf{q})$ is then said to be s-fold degenerate), then the degree of Eq. (A1.76) with respect to ω_p^2 will decrease down to $3N - s + 1$ and we seemingly lose s-1 roots. As a matter of fact, regarding the s-fold degenerate squared frequency as a limiting case of s very close values, it is clear that s-1 "lost" roots are just equal to the degenerate value $\omega_v^2(\mathbf{q})$, since the alternation principle leads them to lie between values tending to each other.

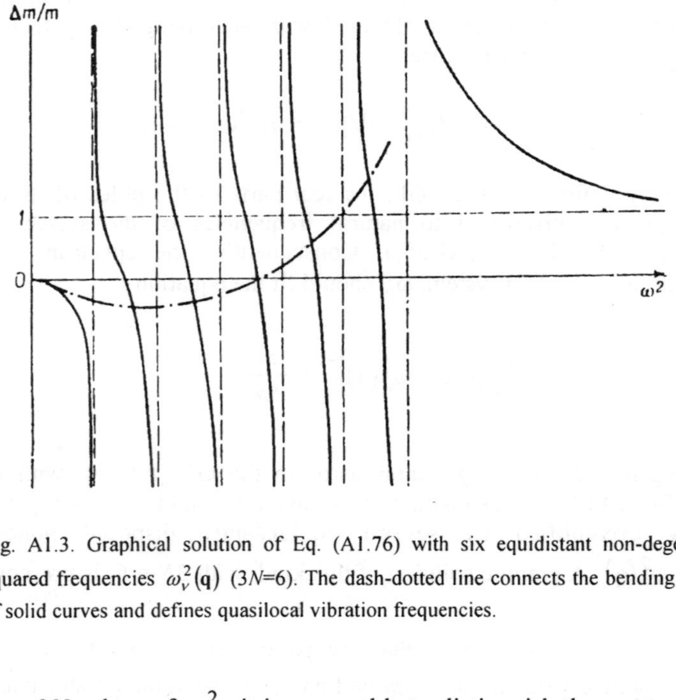

Fig. A1.3. Graphical solution of Eq. (A1.76) with six equidistant non-degenerate squared frequencies $\omega_v^2(\mathbf{q})$ ($3N$=6). The dash-dotted line connects the bending points of solid curves and defines quasilocal vibration frequencies.

Among $3N$ values of ω_p^2, it is reasonable to distinguish the most sensitive to a change in the mass m_C of a "defective" particle. With $m_C < m$, the largest changes are exhibited by the maximum value of ω_p^2 that tends to infinity as $m_C \to 0$ ($m/\Delta m \to +1$). The vibration corresponding to this value ω_p^2 is called the local vibration of a "defective" particle. The implication of the term is that the displacement amplitudes for particles which vibrate with the local frequency ω_p drop fast away from a "defective" site $\mathbf{n} = 0$. Indeed, the analysis of Eqs. (A1.70) and (A1.71) demonstrates that the displacement amplitude $u^{\alpha}(\mathbf{n})$ for a particle at the site $\mathbf{n}$ is related to that for a central particle (of mass m_C), $u^{\alpha}(0)$, by the unperturbed GF :

$$u^{\alpha}(\mathbf{n}) = \Delta m \omega^2 \sum_{\beta} G_{\mathbf{n}0}^{\alpha\beta}(\omega) u^{\beta}(\omega). \tag{A1.77}$$

With $m_C << m$, we have $\omega_p^2 \approx (m/m_C)\langle \omega_\nu^2(\mathbf{q})\rangle >> \max \omega_\nu^2(\mathbf{q})$ and the GF $G_{\mathbf{n}0}^{\alpha\beta}(\omega)$ defined by Eq. (A1.55) decreases fast with increasing $|\mathbf{n}|$. The law governing this decrease is given by a specific form of the functions $\omega_\nu^2(\mathbf{q})$ and $\mathbf{e}_\nu(\mathbf{q})$ but the decrease as it is can easily be noticed even in the limit $G_{\mathbf{n}0}^{\alpha\beta}(\omega) \to \left(m\omega_p^2\right)^{-1} \delta_{\alpha\beta}\delta_{\mathbf{n}0}$.

At $m_C > m$, it is also possible to distinguish the frequencies ω_p most sensitive to a change in m_C. This is less trivial than for local vibrations, because the spectrum ω_p^2 at $m/\Delta m < 0$ has no split-off frequencies and alternates with the spectrum $\omega_\nu^2(\mathbf{q})$ (see Fig A1.3). Transform the sum in Eq. (A1.76) to a form convenient for a subsequent consideration. Moreover, the transformation in question will present an illustrative example for a direct deduction of formula (A1.73) for $G_{00}(\omega)$.

Label all squared frequencies $\omega_\nu^2(\mathbf{q})$ of the unperturbed spectrum in increasing order with the index m: $\omega_\nu^2(\mathbf{q}) \equiv \omega_{0m}^2$, so that $\omega_{0m+1}^2 > \omega_{0m}^2$ (m = 1, 2, ..., max, the subscript 0 indicating values from the unperturbed spectrum). In so doing, note that some values ω_{0m}^2 may be associated with several different vibrations (i.e., several different normal coordinates in Eq. (A1.10)), with the degree of degeneracy for the value ω_{0m}^2 designated by ϑ_m. We also introduce the quantity $\delta\omega_{0m}^2$ for the gap between neighboring squared frequencies: $\delta\omega_{0m}^2 \equiv \omega_{0m+1}^2 - \omega_{0m}^2$. Then an arbitrary value ω_p^2 falling within the interval $[\omega_{0p}^2, \omega_{0p+1}^2]$ may be represented as:

$$\omega_p^2 \equiv \omega_{0p}^2 + z_p \delta\omega_{0p}^2, \tag{A1.78}$$

where z_p is the dimensionless parameter ranging from 0 to 1. With this notation, we rewrite the desired sum in Eq. (A1.76):

$$\frac{1}{3N}\sum_{\mathbf{q}\nu}\frac{1}{\omega_p^2 - \omega_\nu^2(\mathbf{q})} = \frac{1}{3N}\sum_{m}\frac{1}{\omega_{0p}^2 - \omega_{0m}^2 + z_p\delta\omega_{0p}^2} = \frac{1}{3N}\frac{\vartheta_p}{z_p\delta\omega_{0p}^2} +$$

$$+\frac{1}{3N}\sum_{m(\neq p)}\vartheta_m\left[\frac{1}{\omega_{0p}^2-\omega_{0m}^2}-\frac{z_p\delta\omega_0^2}{(\omega_{0p}^2-\omega_{0m}^2+z_p\delta\omega_{0p}^2)(\omega_{0p}^2-\omega_{0m}^2)}\right]$$
$$=\frac{1}{3N}\sum_{m(\neq p)}\frac{\vartheta_m}{\omega_{0p}^2-\omega_{0m}^2}$$
$$+\frac{1}{3N\delta\omega_{0p}^2}\left[\frac{\vartheta_p}{z_p}-z_p\sum_{m(\neq p)}\frac{\vartheta_m}{\left(\frac{\omega_{0p}^2-\omega_{0m}^2}{\delta\omega_{0p}^2}+z_p\right)\frac{\omega_{0p}^2-\omega_{0m}^2}{\delta\omega_{0p}^2}}\right]. \qquad \text{(A1.79)}$$

Consider the bracketed sum. Its terms decrease fast as m moves away from p, which provides a number of simplifications concerning the behavior of the functions ϑ_m and ω_{0m}^2 near $m = p$. First, we set $\vartheta_m = \vartheta_p$ and second, regard the spectrum of ω_{0m}^2 as a equidistant one (these assumptions are quite warranted in a small vicinity far from band boundaries and singular points). The sum concerned then takes the form:

$$\sum_{n\neq 0}\frac{1}{n(n+z_p)}=\frac{1}{z_p}\left(\sum_{n\neq 0}\frac{1}{n}-\sum_{n\neq 0}\frac{1}{n+z_p}\right)=-\frac{1}{z_p}\sum_{n\neq 0}\frac{1}{n+z_p}$$
$$=-\frac{1}{z_p}\left(\sum_{n=-\infty}^{\infty}\frac{1}{n+z_p}-\frac{1}{z_p}\right)=\frac{1}{z_p^2}-\frac{\pi}{z_p}\cot\pi z_p. \qquad \text{(A1.80)}$$

We now note that the distribution function for squared frequencies $\rho\left(\omega_p^2\right)$ (see Eq. (A1.27)) is related to $\delta\omega_{0p}^2$ and ϑ_p as follows:

$$\rho\left(\omega_p^2\right)=\frac{1}{3N}\sum_{\mathbf{q}\nu}\delta\left(\omega^2-\omega_\nu^2(\mathbf{q})\right)=\frac{1}{3N}\sum_m\vartheta_m\delta\left(\omega_{0p}^2+\omega_{0m}^2+z_p\delta\omega_{0p}^2\right)$$
$$\approx\frac{1}{3N}\sum_m\vartheta_m\int_0^1 dz_p\delta\left(\omega_{0p}^2+\omega_{0m}^2+z_p\delta\omega_{0p}^2\right)=\frac{\vartheta_m}{3N\delta\omega_{0p}^2}. \qquad \text{(A1.81)}$$

An approximate smoothing with regard to the gap scales in Eq. (A1.81) is implied by the very definition of the state density involving a δ-function and it becomes exact in the limit $N \to \infty$, $\delta\omega_{0p}^2 \to 0$. The first summand in Eq. (A1.79) is also expressible in terms of $\rho(\omega^2)$:

$$\frac{1}{3N}\sum_{m(\neq p)}\frac{\vartheta_m}{\omega_{0p}^2-\omega_{0m}^2}=\sum_{m(\neq p)}\frac{\vartheta_m}{3N\delta\omega_{0m}^2}\frac{\delta\omega_{0m}^2}{\omega_{0p}^2-\omega_{0m}^2}\approx\int_0^\infty\rho\left(\tilde{\omega}^2\right)\frac{d\tilde{\omega}^2}{\omega_{0p}^2-\tilde{\omega}^2}. \quad \text{(A1.82)}$$

which is in accord with the above-introduced notation (A1.74). Substituting formulae (A1.80) - (A1.82) into Eq. (A1.79) yields the final expression valid for $N \to \infty$:

$$\frac{1}{3N}\sum_{\mathbf{q}\nu}\frac{1}{\omega_p^2-\omega_\nu^2(\mathbf{q})}=P\left(\omega_{0p}^2\right)+\pi\rho\left(\omega_{0p}^2\right)\cot\pi z_p\,. \quad \text{(A1.83)}$$

We can now conveniently select, by substituting Eq. (A1.83) into (A1.76), such a value from the spectrum of ω_p^2 that is the most sensitive to a change in the parameter $\varepsilon = \Delta m/m$. For this purpose, differentiate the displacements z_p with respect to ε:

$$\frac{dz_p}{d\varepsilon}=-\frac{\sin^2\pi z_p}{\pi^2\omega_{0p}^2\rho\left(\omega_{0p}^2\right)}=-\frac{\omega_{0p}^2\rho\left(\omega_{0p}^2\right)}{\left[\varepsilon-\omega_{0p}^2P\left(\omega_{0p}^2\right)\right]^2+\pi^2\omega_{0p}^4\rho^2\left(\omega_{0p}^2\right)} \quad \text{(A1.84)}$$

and require that the absolute magnitude of the derivative $dz_p/d\varepsilon$ be maximum. As evident, this is ensured by the following equivalent equalities:

$$\omega_{0p}^2P\left(\omega_{0p}^2\right)=m/\Delta m, \qquad z_p=1/2\,. \quad \text{(A1.85)}$$

The graphic interpretation of the above conditions consists in selecting the branch of the cotangent curve in Eq. (A1.83) that intersects the straight line $y = m/\Delta m$ at the inflection point (see Fig. A1.3 where the inflection points are connected by a dotted line). For the Debye spectrum, we have

$$P(\omega^2) = \frac{3}{\omega_D^2}\left[\frac{1}{2}\frac{\omega}{\omega_D}\ln\left|\frac{\omega_D+\omega}{\omega_D-\omega}\right| - 1\right]$$
$$= \begin{cases} -3\omega_D^{-2} + 3\omega^2/\omega_D^4, & \omega << \omega_D \\ \omega^{-2} + 3\omega_D^2/5\omega^4, & \omega >> \omega_D \end{cases} \tag{A1.86}$$

and the graphic solution for Eq. (A1.85) is shown in Fig. A1.4.

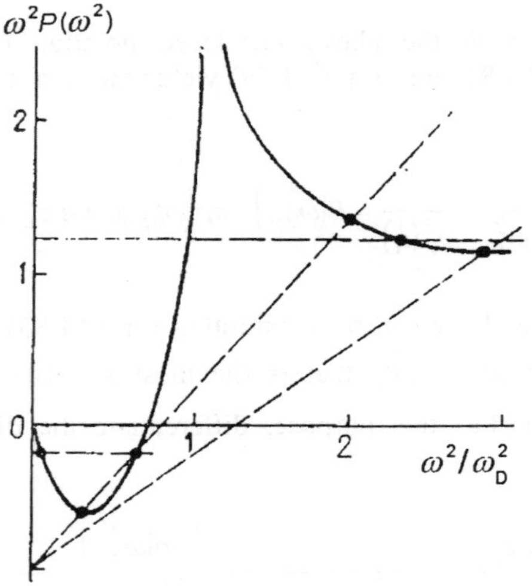

Fig. A1.4. Graphical solution of Eqs. (A1.85) and (A1.99) for a Debye spectrum.

Two roots lie in the range -0.635 < $m/\Delta m$ < 0 (m_C > 2.57 m), the root with a smaller value ω_p^2 corresponding to a smaller value $\rho(\omega_p^2)$, so that it is this root which is the most sensitive to a change in m_C on the strength of the criterion (A1.84). Vibrations with the frequencies meeting the condition (A1.85) are said to be resonance or quasilocal ones. The imaginary part of the GF (A1.75),

$$\mathrm{Im}\,\widetilde{G}_{00}^{\alpha\alpha}(\omega) = -\frac{\pi\rho(\omega^2)}{m}\left[\left(1-\frac{\Delta m}{m}\omega^2P(\omega^2)\right)^2 + \left(\frac{\Delta m}{m}\omega^2\pi\rho(\omega^2)\right)^2\right]^{-1}, \tag{A1.87}$$

which is proportional to the absorbed radiation power for the frequency ω and, in view of the fluctuational-dissipation theorem, to the mean-root-square vibration amplitude for an impurity atom is indeed of the resonance nature.

It should be pointed out that the term with ctgπz_p in Eq. (A1.83) separates out the contribution to the sum that varies fast on a small scale of displacements $\delta\omega_{0p}^2$ (cf. Figs. A1.3 and A1.4). Adding an infinitesimal imaginary quantity to z_p (involved in Eq. (A1.73)) followed by roughening with respect to z_p (as in Eq. (A1.81)) results in cot($\pi z_p + i0$) replaced by -*i*:

$$\int_0^1 \cot(\pi z_p + i0)dz_p = \frac{1}{\pi}\ln\frac{\sin(\pi + i0)}{\sin(i0)} = \frac{1}{\pi}\ln(-1) = -i\,.$$

(The minus sign for *i* in the last equality is chosen to correspond to the sign of the imaginary part ctg($\pi z_p + i0$)). Thus, Eq. (A1.83), if roughened with respect to z_p, turns to Eq. (A1.73). For positive values $m/\Delta m > 1$ ($m_C > m$), we get $\omega_p > \omega_D$, $\rho(\omega_p^2) \to 0$ and return to the local vibrations whose frequencies are specified by split-off branches in Figs. A1.3 and A1.4.

The studies in the effect caused by defects on lattice vibrations were pioneered by Lifshits[139,188]. Much later a related research of applied character was performed by Montroll et al.[189-191]. The discovery of the Mössbauer effect[192] gave impetus to investigating defect effects on dynamic properties of crystals and, most importantly, to predicting resonance vibrations[193-195]. A more detailed account on the theory for local and quasilocal vibrations of a lattice defect can be found in monografs.[196,197] Vibrations of an impurity atom or an impurity with internal degrees of freedom were examined in Refs. 198, 199. In conclusion of the present Appendix, we consider the vibrations of an impurity oscillator incorporated into a system of other bound oscillators.

Assume the impurity particle C to be harmonically bound to a main system of oscillators numbered by i = 0, 1, 2, ... through a single particle with the number i = 0. Fig. A1.5 shows particles labeled by i at cubic lattice sites. The complete Hamiltonian function of the system under discussion is represented as follows:

$$H = \frac{\mathbf{p}_C^2}{2m_C} + \frac{1}{2}\sum_\alpha k_\alpha\left(u_C^\alpha - u_0^\alpha\right)^2 + \sum_{i=0}^{N}\frac{\mathbf{p}_i^2}{2m_i} + \frac{1}{2}\sum_{ij}\Phi_{ij}^{\alpha\beta}u_i^\alpha u_j^\beta - \sum_\alpha F^\alpha(t)\left(u_C^\alpha - u_0^\alpha\right). \tag{A1.88}$$

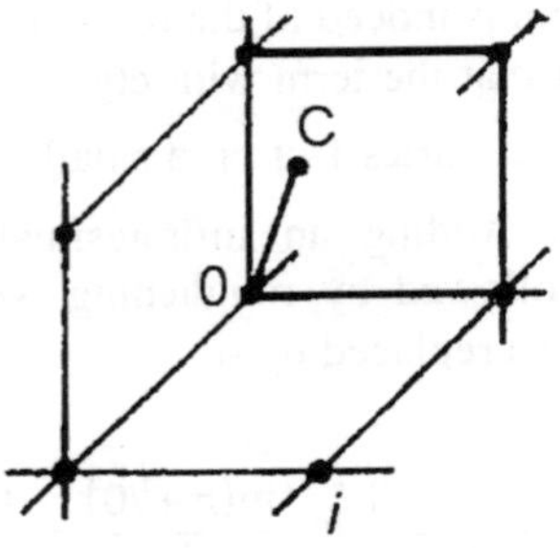

Fig. A1.5. An impurity particle C in a crystal with a cubic lattice.

The last term of the above equation amounts to the work done by the external force $\mathbf{F}(t)$ swaying the impurity bond. The equations of motion for frequency Fourier components of the displacements $\tilde{\mathbf{u}}(\omega)$ and the force $\tilde{\mathbf{F}}(\omega)$ that correspond to Eq. (A1.88) have the form:

$$-m_C\omega^2\tilde{u}_C^\alpha + k_\alpha\left(\tilde{u}_C^\alpha - \tilde{u}_0^\alpha\right) = \tilde{F}^\alpha(\omega), \tag{A1.89}$$

$$-m_i\omega^2\tilde{u}_i^\alpha + \sum_{ij\beta}\Phi_{ij}^{\alpha\beta}\tilde{u}_j^\beta + k_\alpha\left(\tilde{u}_i^\alpha - \tilde{u}_C^\alpha\right)\delta_{i0} = -\tilde{F}^\alpha(\omega)\delta_{i0}. \tag{A1.90}$$

By Eq. (A1.89), we express $\tilde{u}_C^\alpha - \tilde{u}_0^\alpha$ in terms of $\tilde{u}_0^\alpha$:

$$-m_C\omega^2\tilde{u}_C^\alpha + k_\alpha\left(\tilde{u}_C^\alpha - \tilde{u}_0^\alpha\right) = \tilde{F}^\alpha(\omega) \tag{A1.91}$$

and substitute the result into Eq. (A1.90):

$$-m_i\omega^2\tilde{u}_i^\alpha + \sum_{j\beta}\Phi_{ij}^{\alpha\beta}\tilde{u}_j^\beta = -\frac{m_C\omega^2\left(\tilde{F}^\alpha(\omega) + k_\alpha\tilde{u}_C^\alpha\right)}{m_C\omega^2 - k_\alpha}\delta_{i0}. \tag{A1.92}$$

Compare Eq. (A1.92) with the frequency Fourier transform of Eqs. (A1.20) and (A1.21). The right-hand side of Eq. (A1.92) may be regarded as the Fourier component of the external force, but then it turns out that

$$\tilde{u}_i^{\alpha}(\omega) = \sum_{\beta} G_{i0}^{\alpha\beta}(\omega) \frac{m_C \omega^2 \left(\tilde{F}^{\beta}(\omega) + k_{\alpha} \tilde{u}_C^{\beta} \right)}{m_C \omega^2 - k_{\alpha}}, \tag{A1.93}$$

where $G_{i0}^{\alpha\beta}(\omega)$ is the frequency GF of the main oscillator system that is defined by Eq. (A1.23). Rewrite Eq. (A1.93) for $i = 0$ as:

$$\sum_{\beta} \left[\delta_{\alpha\beta} - \frac{m_C \omega^2 k_{\beta}}{m_C \omega^2 - k_{\beta}} G_{00}^{\alpha\beta}(\omega) \right] \tilde{u}_0^{\alpha}(\omega) = \sum_{\beta} \frac{m_C \omega^2}{m_C \omega^2 - k_{\beta}} G_{00}^{\alpha\beta}(\omega) \tilde{F}^{\beta}(\omega) \tag{A1.94}$$

and immediately arrive at the equation for the frequency spectrum ω_p^2 of a system with an impurity oscillator:

$$\det \left[\delta_{\alpha\beta} - \frac{m_C \omega_p^2 k_{\beta}}{m_C \omega_p^2 - k_{\beta}} \operatorname{Re} G_{00}^{\alpha\beta}(\omega_p) \right] = 0 . \tag{A1.95}$$

In the limit $\omega_p^2 >> \omega_{\mathbf{q}\nu_{\max}}^2$, we obtain $G_{00}^{\alpha\beta}(\omega_p) \to \left(m_0 \omega_p^2 \right)^{-1} \delta_{\alpha\beta}$ and Eq. (A1.95) yields, as it must, three roots, $\omega_{p\alpha}^2 = k_{\alpha}/m$, that correspond to the vibrations of oscillators with reduced masses $m \equiv m_C m_0 / (m_C + m_0)$ and force constants k_{α} (along the axes $\alpha = x, y, z$). In this limiting case, $k_{\alpha}/m >> \omega_{\mathbf{q}\nu_{\max}}^2$, but the cofactor for $\operatorname{Re} G_{00}^{\alpha\beta}(\omega_p)$ in Eq. (A1.95) is equal to $-m_C \omega_p^2$ in the frequency range $\omega_p^2 \sim \omega_{\mathbf{q}\nu_{\max}}^2$ and the situation reduces to the above-regarded vibrations of a defect with the mass $m_0 + m_C$ (cf. Eq. (A1.72) where Δm is represented by $-m_C$).

An important comment is relevant here. The degree of Eq. (A1.95) with respect to ω_p^2 is seemingly equal to $3(3N + 1)$, whereas a system with an impurity particle of the mass m_C has only $3N + 3$ degrees of freedom. Indeed, numerators of diagonal elements of the determinant, with the whole sum over $q\nu$ fractions in $\operatorname{Re} G_{00}^{\alpha\beta}(\omega_p)$

(see Eq. (A1.23)) reduced to a common denominator, produce a polynomial of the degree $3N + 1$ in ω_p^2. On the other hand, calculation of the determinant 3×3 will result in a 3 times larger degree of the polynomial. In actual fact, the determinant (A1.95) represents a polynomial of the degree $3N + 3$ in ω_p^2 and has $3N + 3$ roots. To gain a better insight into the case, we make use of the known transformation[139] of the general expression:

$$D(z) = \det\left[\delta_{\alpha\beta} + a_\beta(z) G_{\alpha\beta}(z)\right], \quad G_{\alpha\beta}(z) = \sum \frac{l_q^\alpha l_q^\beta}{\lambda_q - z}, \tag{A1.96}$$

where λ_q with different q has different values. The correspondence between the above relation and Eq. (A1.95) (or the determinant that follows from Eq. (A1.72) and refers to the situation with a mass defect) is quite obvious. Further, Eq. (A1.96) can be written in a detailed form:

$$D(z) = 1 + \sum_\alpha a_\alpha(z) G_{\alpha\alpha}(z) + \frac{1}{2}\sum_{\alpha\beta} \begin{vmatrix} a_\alpha G_{\alpha\alpha} & a_\beta G_{\alpha\beta} \\ a_\alpha G_{\beta\alpha} & a_\beta G_{\beta\beta} \end{vmatrix}$$

$$+ \frac{1}{3!}\sum_{\alpha\beta\gamma} \begin{vmatrix} a_\alpha G_{\alpha\alpha} & a_\beta G_{\alpha\beta} & a_\gamma G_{\alpha\gamma} \\ a_\alpha G_{\beta\alpha} & a_\beta G_{\beta\beta} & a_\gamma G_{\beta\gamma} \\ a_\alpha G_{\gamma\alpha} & a_\beta G_{\gamma\beta} & a_\gamma G_{\gamma\gamma} \end{vmatrix}$$

$$= 1 + \sum_\alpha a_\alpha \sum_q \frac{\left(l_q^\alpha\right)^2}{\lambda_q - z} + \frac{1}{2}\sum_{\alpha\beta} a_\alpha a_\beta \sum_{q_1 q_2} \frac{l_{q_1}^\alpha l_{q_2}^\beta}{(\lambda_{q_1} - z)(\lambda_{q_2} - z)} \begin{vmatrix} l_{q_1}^\alpha & l_{q_2}^\alpha \\ l_{q_1}^\beta & l_{q_2}^\beta \end{vmatrix} \tag{A1.97}$$

$$+ \frac{1}{3!}\sum_{\alpha\beta\gamma} a_\alpha a_\beta a_\gamma \sum_{q_1 q_2 q_3} \frac{l_{q_1}^\alpha l_{q_2}^\beta l_{q_3}^\gamma}{(\lambda_{q_1} - z)(\lambda_{q_2} - z)(\lambda_{q_3} - z)} \begin{vmatrix} l_{q_1}^\alpha & l_{q_2}^\alpha & l_{q_3}^\alpha \\ l_{q_1}^\beta & l_{q_2}^\beta & l_{q_3}^\beta \\ l_{q_1}^\gamma & l_{q_2}^\gamma & l_{q_3}^\gamma \end{vmatrix}$$

With the same values of q_i (i = 1, 2, 3), determinants with l_q^α have the same columns and go to zero, so that no denominator in Eq. (A1.97) can have a cofactor (λ_{qi} - z) to a higher power than the first. Decomposing the expression obtained into simple fractions

$$\frac{1}{(\lambda_{q_1}-z)...(\lambda_{q_n}-z)}=\sum_{i=1}^{n}\left(\prod_{j(\neq i)}\frac{1}{\lambda_{q_j}-\lambda_{q_i}}\right)\frac{1}{\lambda_{q_i}-z}$$

we arrive at:

$$D(z)=1+\sum_q \frac{b_q(z)}{\lambda_q - z},$$

$$b_q(z)=\sum_\alpha a_\alpha(z)\left(l_q^\alpha\right)^2+\sum_{\alpha\beta}a_\alpha(z)a_\alpha(z)l_q^\alpha\sum_{q_1}\frac{l_{q_1}^\beta}{(\lambda_{q_1}-\lambda_q)}\begin{vmatrix} l_q^\alpha & l_{q_1}^\alpha \\ l_q^\beta & l_{q_1}^\beta \end{vmatrix} \tag{A1.98}$$

$$+\frac{1}{2}\sum_{\alpha\beta\gamma}a_\alpha a_\beta a_\gamma l_q^\alpha \sum_{q_1 q_2}\frac{l_{q_1}^\beta l_{q_2}^\gamma}{(\lambda_{q_1}-\lambda_q)(\lambda_{q_2}-\lambda_q)}\begin{vmatrix} l_q^\alpha & l_{q_1}^\alpha & l_{q_2}^\alpha \\ l_q^\beta & l_{q_1}^\beta & l_{q_2}^\beta \\ l_q^\gamma & l_{q_1}^\gamma & l_{q_2}^\gamma \end{vmatrix}.$$

In the problem on a mass defect, a_α is not a function of z, and q takes on 3N values. Therefore, on reducing the expression (A1.98) to a common denominator, the numerator $D(z)$ becomes a polynomial in z of the degree $3N$. In the problem on an impurity oscillator, $a_\alpha(z)$ also has a denominator of the form $(\lambda_\alpha - z)$ but differs in that the nominator $D(z)$ is a polynomial of the degree $3N + 3$, as we set out to prove.

For an isotropic lattice, we take advantage of the transform given in Eqs. (A1.73) and (A1.74) to derive, from Eq. (A1.95), three equations in frequencies of local and quasilocal vibrations of an impurity oscillator:

$$\omega_p^2 P\left(\omega_p^2\right)=\frac{m_0}{m_C}\frac{\omega_p^2-\omega_\alpha^2}{\omega_\alpha^2}, \quad \omega_\alpha^2 \equiv \frac{k_\alpha}{m_C}, \quad \alpha = x,\, y,\, z\,. \tag{A1.99}$$

The distinction of the equation (A1.99) from Eq. (A1.85) consists in the linear dependence of the right-hand side on ω_p^2. Let us analyze the solutions of Eqs. (A1.99) for the Debye spectrum (A1.65), (A1.67), and (A1.86). Fig. A1.4 illustrates that local vibrations occur at $\omega_\alpha \geq \omega_D$ and $m_C \leq m_0$. Quasilocal vibrations arise at $\omega_\alpha \leq \omega_D$ and $m_C \leq m_0$. It is noteworthy that the Debye approximation is not valid for frequencies close to ω_D and the graphic analysis given in Fig. A1.4 is hence meaningful only for roots ω_p^2 distant enough from ω_D.

In the case of an isotropic lattice, Eq. (A1.94) is readily solved for $\tilde{u}_0^\alpha(\omega)$:

$$\tilde{u}_0^\alpha(\omega)=\frac{1}{m_C\omega_\alpha^2}\,\frac{\omega^2 P(\omega^2)-i\pi\omega^2\rho(\omega^2)}{\dfrac{m_0}{m_C}\dfrac{\omega^2-\omega_\alpha^2}{\omega_\alpha^2}-\omega^2 P(\omega^2)+i\pi\omega^2\rho(\omega^2)}\,\tilde{F}^\alpha(\omega). \tag{A1.100}$$

Substituting Eq. (A1.100) into (A1.91), we find:

$$\tilde{u}_C^\alpha(\omega)-\tilde{u}_0^\alpha(\omega)=\chi^\alpha(\omega)\,\tilde{F}^\alpha(\omega)\,, \tag{A1.101}$$

where $\chi^\alpha(\omega)$ is the impurity bond susceptibility equal to:

$$\chi^\alpha(\omega)=-\frac{1}{m_C\omega_\alpha^2}\,\frac{\dfrac{m_0}{m_C}+\omega^2 P(\omega^2)-i\pi\omega^2\rho(\omega^2)}{\dfrac{m_0}{m_C}\dfrac{\omega^2-\omega_\alpha^2}{\omega_\alpha^2}-\omega^2 P(\omega^2)+i\pi\omega^2\rho(\omega^2)}\,. \tag{A1.102}$$

The imaginary part of $\chi^\alpha(\omega)$ proportional to the absorption spectrum for impurity bond vibrations is specified by the relation:

$$\operatorname{Im}\chi^\alpha(\omega)=-\frac{m_0\omega^2}{m_C^2\omega_\alpha^4}\,\frac{\pi\omega^2\rho(\omega^2)}{\left[\dfrac{m_0}{m_C}\dfrac{\omega^2-\omega_\alpha^2}{\omega_\alpha^2}-\omega^2 P(\omega^2)\right]^2+\left[\pi\omega^2\rho(\omega^2)\right]^2}\,. \tag{A1.103}$$

In the frequency range for local vibrations, where $\rho(\omega^2)\to\infty$, the function Im $\chi^\alpha(\omega)$ represents an infinitely narrow and high spike at $\omega^2=\omega_p^2$ (ω_p^2 is the root of Eq. (A1.99)). In the frequency range for quasilocal vibrations, the function (A1.103) has a resonance character with a finite half-width.

Introduce the spectral function normalized to unity:

$$S(\omega^2)=\frac{C}{\pi}\operatorname{Im}\chi^\alpha(\omega),\qquad \int_0^\infty S(\omega^2)\,d\omega^2=1\,. \tag{A1.104}$$

The normalization constant C can be found by calculating the zero moment of the distribution $S(\omega^2)$. Indeed, since the GF of an impurity bond amounts to $-\chi^\alpha(\omega)$, it follows from formulae (A1.74), (A1.67), and (A1.102) that

$$1 = M_0 = -\lim_{\omega\to\infty} \omega^2 C\chi^\alpha(\omega) = \left(\frac{1}{m_0} + \frac{1}{m_C}\right)C \equiv \frac{C}{m} \tag{A1.105}$$

and hence $C = m = m_C m_0/(m_C + m_0)$.

Assume Eq. (A1.99) to have a single root ω_p^2. Then the function (A1.104), with formally set $\rho(\omega^2) \to 0$, should become $S(\omega^2) = \delta(\omega^2 - \omega_p^2)$. Furthermore, in this limit,

$$S(\omega^2) = \frac{m m_0 \omega^2}{m_C \omega_\alpha^4}\delta\left[\frac{m_0}{m_C}\frac{\omega^2 - \omega_\alpha^2}{\omega_\alpha^2} - \omega^2 P(\omega^2)\right]$$

$$= \frac{m_0^2\omega^2}{m_C(m_0 + m_C)\omega_\alpha^4}\left[\frac{d}{d\omega^2}\left(\frac{m_0}{m_C}\frac{\omega^2 - \omega_\alpha^2}{\omega_\alpha^2} - \omega^2 P(\omega^2)\right)\right]^{-1}_{\omega^2 = \omega_p^2} \delta(\omega^2 - \omega_p^2),$$

so that

$$\left[\frac{d}{d\omega^2}\left(\frac{m_0}{m_C}\frac{\omega^2 - \omega_\alpha^2}{\omega_\alpha^2} - \omega^2 P(\omega^2)\right)\right]_{\omega^2 = \omega_p^2} = \frac{m_0^2\omega_p^2}{m_C(m_0 + m_C)\omega_\alpha^4}. \tag{A1.106}$$

By approximately expanding the term vanishing at $\omega^2 = \omega_p^2$ in the denominator of Eq. (A1.103), we obtain the Lorentz approximation for $S(\omega^2)$:

$$S(\omega^2) \approx \frac{1}{\pi}\frac{m_C(m_0 + m_C)\omega_\alpha^4 \pi\rho(\omega_p^2)/m_0^2}{(\omega^2 - \omega_0^2)^2 + \left[m_C(m_0 + m_C)\omega_\alpha^4 \pi\rho(\omega_p^2)/m_0^2\right]^2}. \tag{A1.107}$$

This yields the halfwidth of the distribution maximum for $S(\omega^2)$:

$$\Delta\omega_{1/2}^2 \approx \frac{m_C(m_0 + m_C)}{m_0^2}\omega_\alpha^4 \pi\rho(\omega_p^2). \tag{A1.108}$$

Spectral line halfwidths measured over the scale of frequencies rather than squared frequencies are clearly related to: $\Delta\omega_{1/2} \approx \Delta\omega_{1/2}^2 / 2\omega_p$. Substituting here the explicitly expressed quantity $\rho(\omega_p^2)$ for the Debye spectrum (see Eqs. (A1.65) and (A1.67)), we are led to:

$$\Delta\omega_{1/2} \approx \frac{3\pi}{4}\frac{m_C(m_0 + m_C)}{m_0^2}\left(\frac{\omega_\alpha}{\omega_D}\right)^3 \omega_\alpha = \frac{m_0 + m_C}{m_0}\frac{m_C\omega_\alpha^4}{8\pi\rho\bar{c}^3}, \tag{A1.109}$$

where $\rho = m_0/V_0$ is the crystal density.

In normalizing the expression (A1.103), we have already taken advantage of the fact that the quantity $-\chi^\alpha(\omega)$ is the GF $G^{\alpha\alpha}(\omega)$ of an impurity bond. It can be strictly defined in a standard manner:

$$G^{\alpha\beta}(t) = -\frac{i}{\hbar}\theta(t)\left\langle\left[u_C^\alpha(t) - u_0^\alpha(t),\ u_C^\alpha(0) - u_0^\alpha(0)\right]\right\rangle. \tag{A1.110}$$

Setting up the equations of motion for GF $G^{\alpha\beta}(t)$ with the Hamiltonian of the problem (A1.88), we can relate its frequency Fourier component with the GF $G_{00}^{\alpha\beta}(\omega)$ for the main subsystem of particles. Actually, this idea has already been implemented in terms of Fourier components of displacements $\tilde{u}_C^\alpha(\omega)$ and $\tilde{u}_0^\alpha(\omega)$ (see Eqs. (A1.89) - (A1.94) and (A1.100) - (A1.102)). Yet, after writing Eq. (A1.94), we discussed the case of an isotropic lattice and applied the transform (A1.73). Let us consider the result of a more general treatment in which the GF $G_{00}^{\alpha\beta}(\omega)$ is diagonal but not necessarily isotropic (as for Rayleigh surface vibrations). In this case, the relationship between $G^{\alpha\beta}(\omega)$ and $G_{00}^{\alpha\beta}(\omega)$ is given by the formula

$$G^{\alpha\alpha}(\omega) = \frac{1}{m_C\omega_\alpha^4}\frac{1 + m_C\omega^2 G_{00}^{\alpha\alpha}(\omega)}{(\omega^2 - \omega_\alpha^2)/\omega_\alpha^2 - m_C\omega^2 G_{00}^{\alpha\alpha}(\omega)} \tag{A1.111}$$

that generalizes Eq. (A1.102).

Given two identical impurity particles bound to the atoms $i = 0$ and $i = 1$, and provided external forces $F^{\alpha}(t)$ are absent, the system of equations (A1.89) - (A1.90) is generalized as follows:

$$\begin{cases} -m_C\omega^2\tilde{u}_{C0}^{\alpha} + k_{\alpha}(\tilde{u}_{C0}^{\alpha} - \tilde{u}_0^{\alpha}) = 0, \\ -m_C\omega^2\tilde{u}_{C\mathbf{n}}^{\alpha} + k_{\alpha}(\tilde{u}_{C\mathbf{n}}^{\alpha} - \tilde{u}_{\mathbf{n}}^{\alpha}) = 0, \\ -m_i\omega^2\tilde{u}_i^{\alpha} + \sum_{j\beta}\Phi_{ij}^{\alpha\beta}\tilde{u}_j^{\alpha} + k_{\alpha}(\tilde{u}_0^{\alpha} - \tilde{u}_{C0}^{\alpha})\delta_{i0} + k_{\alpha}(\tilde{u}_{\mathbf{n}}^{\alpha} - \tilde{u}_{C\mathbf{n}}^{\alpha})\delta_{j\mathbf{n}} = 0. \end{cases} \tag{A1.112}$$

Performing transformations like (A1.91) - (A1.94), we derive the system of equations:

$$\begin{cases} \sum_{\beta}\left\{\left[\delta_{\alpha\beta} - \dfrac{m_C\omega^2 k_{\beta}}{m_C\omega^2 - k_{\beta}}G_{00}^{\alpha\beta}(\omega)\right]\tilde{u}_0^{\beta}(\omega) - \dfrac{m_C\omega^2 k_{\beta}}{m_C\omega^2 - k_{\beta}}G_{0\mathbf{n}}^{\alpha\beta}(\omega)\tilde{u}_{\mathbf{n}}^{\beta}(\omega)\right\} = 0, \\ \sum_{\beta}\left\{-\dfrac{m_C\omega^2 k_{\beta}}{m_C\omega^2 - k_{\beta}}G_{\mathbf{n}0}^{\alpha\beta}(\omega)\tilde{u}_0^{\beta}(\omega) + \left[\delta_{\alpha\beta} - \dfrac{m_C\omega^2 k_{\beta}}{m_C\omega^2 - k_{\beta}}G_{\mathbf{nn}}^{\alpha\beta}(\omega)\right]\tilde{u}_{\mathbf{n}}^{\beta}(\omega)\right\} = 0. \end{cases} \tag{A1.113}$$

For a monoatomic lattice treated in the approximation of isotropic GFs, the equation is deduced for natural frequencies of the system:

$$\begin{gathered} \frac{\omega^2 - \omega_C^2}{\omega_C^2} = m_C\omega^2\,\mathrm{Re}\left[G(0,\omega) - G(\mathbf{n},\omega)\right], \\ G(\mathbf{n},\omega) = \frac{1}{mN}\sum_{\mathbf{q}}\frac{e^{i\mathbf{q}\cdot\mathbf{n}}}{\omega^2 - \omega^2(\mathbf{q}) + i0\,\mathrm{sign}\,\omega}. \end{gathered} \tag{A1.114}$$

where $\omega_C^2 = k/m_C$. As pointed out above, the GF $G(\mathbf{n},\omega)$ decreases fast with increasing $|\mathbf{n}|$. Importantly, the explicit form of the drop is much dependent on the form of the function $\omega(\mathbf{q})$ not only at small values q, but also near the boundary of the first Brillouin zone. To estimate frequency displacements induced by crystal-mediated (i.e., indirect) interaction between impurity particles, we restrict ourselves to the case of a linear monoatomic chain which enables, due to the simple explicit form of the dispersion law ($\omega(q) = \omega_D|\sin(qa/2)|$ with a denoting the lattice constant) and also due to the linearity of all displacements (justifying a index-free notation in Eq. (A1.114)), the value $G(n, \omega)$ to be conveniently calculated at $\omega > \omega_D$:

$$G(n, \omega) = \frac{a}{\pi m} \int_0^{\pi/a} \frac{\cos qna}{\omega^2 - \omega_D^2 \sin^2(qa/2)} dq = \frac{1}{\pi m} \int_0^{\pi} \frac{\cos nx}{\omega^2 - \frac{1}{2}\omega_D^2 + \frac{1}{2}\omega_D^2 \cos x} dx$$

$$= \frac{(-1)^n}{m\omega\sqrt{\omega^2 - \omega_D^2}} \left(\frac{\omega^2 - \omega_D^2/2 - \omega\sqrt{\omega^2 - \omega_D^2}}{\omega_D^2/2} \right)^n . \tag{A1.115}$$

The parenthesized expression is less than unity and hence $G(n, \omega)$ decreases exponentially with increasing n. The asymptotic (A1.115) at $\omega >> \omega_D$ appears as

$$G(n, \omega) \approx \frac{(-1)^n}{m\omega^2} \left(1 + \frac{1}{2} \frac{\omega_D^2}{\omega^2} \right) \left(\frac{\omega_D^2}{4\omega^2} \right)^n . \tag{A1.116}$$

Therefore, in the case $\omega >> \omega_D$, local vibration frequencies are approximately equal to:

$$\omega^2 \approx \left[1 + \frac{1}{2} \frac{m_C}{m + m_C} \frac{\omega_D^2}{\tilde{\omega}_C^2} \pm \frac{m_C}{m + m_C} \left(\frac{\omega_D^2}{4\tilde{\omega}_C^2} \right)^n \right] \tilde{\omega}_C^2 , \tag{A1.117}$$

where $\tilde{\omega}_C^2 \equiv \omega_C^2 (m + m_C)/m = k(m + m_C)/mm_C$ is the squared frequency of an oscillator with the force constant k and the reduced mass $mm_C/(m + m_C)$. The second term in Eq. (A1.117) accounts for a difference between the local vibration frequency for an isolated impurity particle and the value $\tilde{\omega}_C^2$. The third term represents the contribution from the interaction of impurity particles separated by the distance na, the two signs corresponding to their in-phase and anti-phase vibrations.

Appendix 2

Thermally activated reorientations and tunnel relaxation of orientational states in a phonon field

The rotational mobility of adsorbed molecules is caused by its rotational degree of freedom (resulting from the fact that the molecule is tightly bound to the substrate through the only atom) and by the coupling of molecular vibrations with surface atomic vibrations. The rotational motion intensity is strongly temperature-dependent and affects spectroscopic characteristics. As a result, the rotational mobility of surface hydroxyl groups was reliably detected.[200-203]

Consider reorientations of a diatomic surface group BC (see Fig. A2.1) connected to the substrate thermostat. By a reorientation is meant a transition of the atom C from one to another well of the azimuthal potential $U(\varphi)$ (see Fig. 4.4)). The terminology used implies a classical (or at least quasi-classical) description of azimuthal motion allowing the localization of the atom C in a certain well. A classical particle, with the energy lower than the reorientation barrier ΔU_φ, which does not interact with the thermostat cannot leave the potential well where it was located initially. The only pathway to reorientations is provided by energy fluctuations of a particle which arise from its contact with the thermostat. Let us estimate the average frequency of reorientations in the framework of this classical approach.

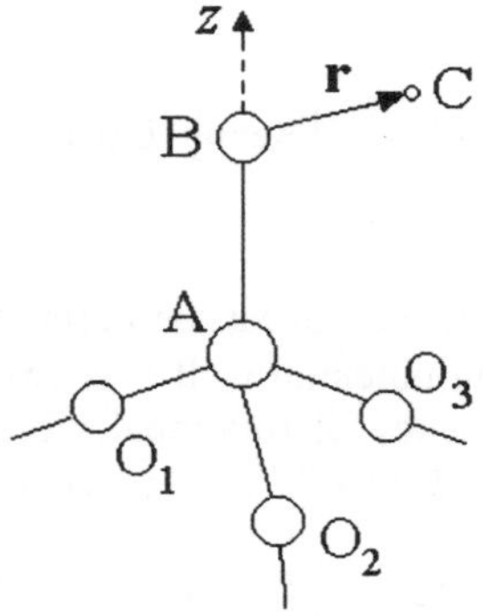

Fig. A2.1. Schematic depiction of a diatomic surface group BC.

Let the minimum of the potential $U(\varphi)$ be characterized by the angular coordinate value $\varphi = 0$ and two nearest maxima be at the points $\varphi = \pm\,\varphi_m$. The probability density $f(\omega, \varphi)$ for a particle to be located at a point with the angular coordinate φ and to have the angular velocity ω under thermodynamic equilibrium with a thermostat is given by the Gibbs distribution:

$$f(\omega,\varphi)=\frac{\exp(-H/k_BT)}{\int\limits_{-\infty}^{\infty} d\omega \int\limits_{-\varphi m}^{\varphi m} d\varphi \exp(-H/k_BT)}; \qquad H=\frac{I\omega^2}{2}+U(\varphi), \tag{A2.1}$$

where I is the moment of inertia relative to the rotation axis. During the time dt, the barrier can be surmounted at the point $\varphi = \varphi_m$ only by the particles that fall within the region φ_m - $\omega\,dt < \varphi < \varphi_m$ and move towards the barrier $(\omega > 0)$. The average number of the jumps over the barrier for the time dt is therefore equal to:

$$dW = dt\int\limits_0^{\infty} d\omega\omega f(\omega,\varphi_m) = dt\frac{\exp[-U(\varphi_m)/k_BT]}{\int\limits_{-\varphi_m}^{\varphi_m} d\varphi \exp[-U(\varphi)/k_BT]}\left(\frac{k_BT}{2\pi I}\right)^{1/2}. \tag{A2.2}$$

If the reorientation barrier is $U(\varphi_m) - U(0) \equiv \Delta U(\varphi) >> k_BT$, then the potential energy $U(\varphi_m)$ in the exponent of the integrand can be expanded in φ accurate to φ^2 and the integral is easily calculable:

$$\int\limits_{-\varphi_m}^{\varphi_m} d\varphi \exp[-U(\varphi)/k_BT] \approx \exp[-U(0)/k_BT]\left(\frac{2\pi k_BT}{U''(0)}\right)^{1/2}. \tag{A2.3}$$

The average reorientation frequency ω can be defined as an average velocity with which a particle leaves a given potential well over two barriers with $\varphi = \pm\,\varphi_m$: $w \equiv 2dW/dt$. With the cyclic frequency of rotational vibrations in the well defined by Eqs. (A2.2) and (A2.3) as $\omega_\varphi = [U''(0)/I]^{1/2}$, we arrive at:

$$w = \pi^{-1}\omega_\varphi \exp\left(-\Delta U_\varphi / k_BT\right). \tag{A2.4}$$

This expression offers a clear physical interpretation. The preexponential is equal to the frequency of particle's jumps onto the well walls, while the exponent represents the probability for the potential well ΔU_φ to be surmounted on thermal activation.

We demonstrate that the spectral function of valence harmonic vibrations of a diatomic group that effects rotational reorientations is broadened by w. The vector of atom C displacements relative to the atom B (see Fig. A2.1) may be represented as $x(t)\mathbf{e}(t)$, where $x(t)$ is the change in the length of the valence bond oriented at the time t along the unit vector $\mathbf{e}(t)$. Characteristic periods of valence vibrations are much shorter than periods of changes in unit vector orientations. As a consequence, the GF of the displacements defined by Eq. (4.2.1) can be expressed approximately as:

$$G^{\alpha\beta}(t) \approx -\frac{i}{\hbar}\theta(t)\langle[x(t),x(0)]\rangle\langle e^{\alpha}(t)e^{\beta}(0)\rangle = G_{xx}(t)\langle e^{\alpha}(t)e^{\beta}(0)\rangle . \qquad \text{(A2.5)}$$

Here we first calculate the commutator $[x(t),x(0)]$ averaged over an equilibrium Gibbs state ensemble with fixed orientations $\mathbf{e}$ so as to obtain the GF of a harmonic oscillator:

$$G_{xx}(t) = -(m\omega_0)^{-1}\theta(t)\sin\omega_0 t \qquad \text{(A2.6)}$$

and then average it over the unit vectors $\mathbf{e}(t)$.

Without regard for deformational and rotational vibrations of unit vectors $\mathbf{e}(t)$, the qualitative behavior of the time dependence of the correlation function for two-dimensional reorientations is describable by the following relation:

$$\langle e^{\alpha}(t)e^{\beta}(0)\rangle = \frac{1}{2}\delta_{\alpha\beta}\exp(-wt),\ t \geq 0 . \qquad \text{(A2.7)}$$

For the initial time $t = 0$, the above formula (A2.7) is identical with the result of averaging over random orientations in a surface plane. In the course of time, the "memory" of the initial orientation fades, the condition $t >> w^{-1}$ (w^{-1} is the average period between reorientations) permitting an independent averaging over $e^{\alpha}(t)$ and $e^{\beta}(0)$, and the correlation function (A2.7) tends to zero.

We substitute (A2.6) and (A2.7) into (A2.5) to obtain the frequency dependence of the spectral function:

$$S^{\alpha\beta}\left(\omega^2\right) = -\frac{1}{\pi}\operatorname{Im}G^{\alpha\beta}(\omega) = \frac{\delta_{\alpha\beta}}{2\pi m\omega_0}\int_0^\infty dt e^{-\omega t}\sin\omega_0 t\sin\omega t$$
$$= \frac{\delta_{\alpha\beta}}{\pi m}\frac{\omega w}{\left(\omega^2-\omega_0^2-w^2\right)^2+4\omega^2 w^2}. \tag{A2.8}$$

It is easy to verify that $S^{\alpha\beta}(\omega^2)$ is normalized by the condition:

$$\int_{-\infty}^{\infty} S^{\alpha\beta}\left(\omega^2\right)\omega\, d\omega = \frac{1}{2m}\delta_{\alpha\beta}. \tag{A2.9}$$

Since the average reorientation frequency w is far less than the cyclic frequency of valence vibrations ω_0, expression (A2.8), with measured frequency values ω close to ω_0, can approximately be rewritten as:

$$S^{\alpha\beta}\left(\omega^2\right) \approx \frac{\delta_{\alpha\beta}}{4\pi m\omega_0}\frac{w}{\left(\omega-\omega_0\right)^2+w^2} \tag{A2.10}$$

(the normalization condition (A2.10) differs from (A2.9) in that it has no the cofactor ω in the integrand but has an additional cofactor $(2\omega_0)^{-1}$ in the right-hand side).

The spectral function thus has a Lorentz shape with a halfwidth at the half distribution height equal to the average reorientation frequency w. If expressed in spectroscopic units (cm^{-1}), the halfwidth $\Delta\nu_{1/2}$ amounts to $\omega/2\pi c_0$ (c_0 designates the velocity of light in vacuo).

Ryason and Russel measured the temperature dependence of the IR absorption band halfwidth for valence vibrations of hydroxyl groups on the silica surface.[200] At $T \geq 325$ K, the least squares method permits a straight line to be drawn through experimental points of the dependence ln $\Delta\nu_{1/2}$ (T^{-1}), the equation of the line appearing as follows:[200]

$$\ln(2\Delta\nu_{1/2}) = -(4.42\pm0.35)\cdot10^2 T^{-1} + (2.79\pm0.08). \tag{A2.11}$$

Hence, the activation energy, $\Delta\varepsilon_{\mathrm{rot}}$, amounts to 38±3 meV and the preexponential, $2\Delta\nu_{1/2}$ ($T\to\infty$) is equal to 16.3±1.3 cm^{-1}. Compare this result with the theoretical estimate (A2.4). If $\Delta\varepsilon_{\mathrm{rot}}$ is identified with the reorientation barrier ΔU_φ, then $\Delta\nu_{1/2}$

$(T\to\infty) = \omega_\varphi/2\pi^2 c_0 = 3(\Delta U_\varphi/2I)^{1/2}/2\pi^2 c_0 \approx 73$ cm^{-1} (with $I = 1.48\cdot10^{-48}$ g·cm^2 for a hydroxyl group), this value being nine times as large as the experimental one. At $T = 300$ K, we have from Eq. (A2.4) that $\Delta\nu_{1/2} \approx 16.8$ cm^{-1}, whereas experiments yield $\Delta\nu_{1/2} \approx 2$ cm^{-1}.

Nevertheless, relation (A2.4) agrees well with experimental characteristics of the Brownian rotational motion of molecules in solid state, when molecular reorientation barriers are several times higher.[159, 204] It comes as no surprise, because the greater are values ΔU_φ, the more applicable is the classical description for reorientations. The criterion here may be represented by the inequality $p \equiv \Delta U_\varphi/\hbar\omega_\varphi = (2I\Delta U_\varphi)^{1/2}/n\hbar >> 1$ implying that the height of the barrier ΔU_φ should include a great number of quanta $\hbar\omega_\varphi$ for rotational vibrations. In this case, it is possible to disregard the discreteness of energy portions derived from the thermostat which are needed for a particle to fluctuationally surmount the barrier ΔU_φ.

For a hydroxyl group on the silica surface, the reduced barrier p is of the order of unity and hence a classical description is no longer applicable. We have already ascertained this by analyzing the empirical relation (A2.11). To refine the estimated value of the energy barrier ΔU_φ,[200] it was taken into account[205] that the activation energy $\Delta\varepsilon_{rot}$ should be reckoned not from the well bottom ($\Delta\varepsilon_{rot}$ would then be equal to ΔU_φ) but from the ground energy level for zero vibrations $\hbar\omega_\varphi/2$. Since ω_φ is dependent on ΔU_φ, we are led to the following quadratic equation for the reorientation barrier ΔU_φ:

$$\Delta E_{rot} = \Delta U_\varphi - 3\hbar\left(\Delta U_\varphi / 2I\right)^{1/2}/2 . \tag{A2.12}$$

With $\Delta\varepsilon_{rot} \approx 38$ meV and $I \approx 1.48\cdot10^{-40}$ g · cm^2, we obtain $\Delta U_\varphi \approx 55$ meV, in agreement with the quantum chemical calculation.[206] However, the relation (A2.4) remains inapplicable, as the preexponential in it proves to be anomalously high.

It is necessary to take proper account of the discreteness of energies transferred to a surface group from the substrate thermostat. If $p \sim 1$, then the first excited level with the energy $\sim 3\hbar\omega_\varphi/2$ lies near the potential well top and the quantum transition to it, when activated by the interaction with the substrate phonon thermostat, will enable the atom C to pass freely over the barrier or under a low barrier by tunneling. In this case, the rate of transitions from the ground to the first excited level is expected to be a good estimate for an average reorientation frequency.

The probability for a transition to occur between two states per unit time is determined by Fermi's golden rule and depends on the operator of interaction between the subsystem concerned and a thermostat. As orientational states are characterized by a low-energy spectrum, they will be substantially influenced by the interaction with low-frequency vibrational modes of a solid-state matrix. In the isotropic elastic continuum approximation, this interaction is expressible as:[207]

$$H_{\text{int}} = \Gamma_{\alpha\beta}(\mathbf{r})u_{\alpha\beta} + m_C\mathbf{r}\cdot\ddot{u}\,, \tag{A2.13}$$

where $u_{\alpha\beta}$ is the deformation tensor, $\Gamma_{\alpha\beta}(\mathbf{r})$ represents the deformation potential,[208] and $\mathbf{r} = \mathbf{r}_C - \mathbf{r}_B$ denotes the orientation vector for the group BC whose center-of-mass vibrations are described by the deformation vector $\mathbf{u}$. In what follows, we shall derive a relation like Eq. (A2.13) and for now it is reasonable to elucidate the physical meaning of its terms. The first term describes the change in the energy of the interaction between the atom C and the substrate induced by the substrate deformation. Its order of magnitude may be estimated by the total energy ε_C of the bonds between the atom C and all nearest-neighboring atoms except the atom B; ε_C being of the same order as reorientation barriers ΔU_φ. The second term allows for the d'Alembert force, $m_C\ddot{\mathbf{u}}$, in a noninertial system of reference connected with the center of mass for the group BC which undergoes acceleration $\ddot{\mathbf{u}}$ as a solid vibrates. If we designate the energy binding the atom B to a solid-state matrix by ε_B and the interatomic distance by a, then we have $|\ddot{\mathbf{u}}| \sim \varepsilon_B/m_B a$ and the second term in Eq. (A2.13) at $r \sim a$ turns out to be of the order $(m_C/m_B)\varepsilon_B$. Thus, the ratio of the second to the first term is estimated as $(m_C/m_B)\cdot(\varepsilon_B/\varepsilon_C)$.[164]

With the atom C strongly bound not only to B but also to the other atoms of a solid-state matrix (i.e., when $\varepsilon_C \sim \varepsilon_B$) the above ratio is small in the parameter $m_C/m_B \ll 1$, so that the dominant contribution to the interaction with phonons is provided by the deformation potential. Reorientation probabilities were calculated, with the deformation term only taken into consideration, in Refs. 209, 210. For a diatomic group BC, $\varepsilon_C \sim \Delta U_\varphi \sim 0.1$ eV, whereas $\varepsilon_B \sim 10$ eV (a typical bond energy for ionic and covalent crystals). A strong binding of the atom C only to the atom B results in the dominant contribution from inertial forces.[211] For OH groups, as an example, the second term in Eq. (A2.13) is more than 6 times as large as the first one.

Calculate the probabilities for the transitions between orientational states $|\alpha\rangle$ of the BC group which are dictated by the d'Alembert force. If the group BC is incorporated into the solid-state matrix bulk (as for instance, hydroxyl groups contained in many biological macromolecules[212] and amorphous substances[213]), then the deformation vector $\mathbf{u}$ is describable by Eq. (A1.43) and the second term in Eq. (A2.13) may, in terms of secondary quantization, be written as:

$$H_{\text{int}}^{(1)} = \sum_{qv}\left(\chi_{qv}b_{qv} + \chi_{qv}^{+}b_{qv}^{+}\right). \tag{A2.14}$$

Here $b_{\mathbf{q}\nu}^{+}$ and $b_{\mathbf{q}\nu}$ are creation and destruction operators for phonons of the *νth* acoustic branch (one longitudinal with $\nu = l$ and two transverse ones with $\nu = t$) with the wave vector $\mathbf{q}$ which obey the dispersion laws $\omega_\nu(\mathbf{q}) = c_\nu|\mathbf{q}|$ (c_l and c_t are the longitudinal and transverse sound velocities). The operator $\chi_{\mathbf{q}\nu}$ is defined by the relation

$$\chi_{q\nu} = -\left(\frac{\hbar m_c^2 \omega_\nu^3(\mathbf{q})}{2\rho V}\right)^{1/2} \mathbf{r}\cdot\mathbf{e}_\nu(\mathbf{q}), \tag{A2.15}$$

where ρ is the medium density, V is the volume of the main area, and $\mathbf{e}_\nu(\mathbf{q})$ is the phonon polarization unit vector. Provided the group BC is located on a continuum surface, the deformation vector $\mathbf{u}$ is given by Eq. (2.35) for Rayleigh phonons and the interaction operator has the form (A2.14), with the summation performed only over the two-dimensional wave vector $\mathbf{q}$ (the index ν is absent) and the operator χ_q specified as follows:

$$\chi_q = -\left(\frac{\xi^4}{4\lambda}\frac{\hbar m_c^2 \omega_p^3(q) q}{2\rho S}\right)^{1/2}\left(i\mathbf{r}_{\parallel}\cdot\frac{\mathbf{q}}{q} - \sqrt{\frac{\kappa_1}{2}}r_z\right). \tag{A2.16}$$

The dispersion law for Rayleigh phonons is specified by the relation $\omega_p(\mathbf{q}) = c_p|\mathbf{q}|$ (c_p is the propagation velocity for Rayleigh waves); $\xi \equiv c_p/c_t$ is the positive root of the following equation:

$$4\left[\left(1-\xi^2\right)\left(1-\xi^2\gamma^2\right)\right]^{1/2} = \left(2-\xi^2\right)^2, \quad \gamma \equiv c_t/c_l\,; \tag{A2.17}$$

the values λ and κ_1 are found from formulae:

$$\lambda = \frac{\xi^4}{4\sqrt{1-\xi^2}}\left[1+\frac{\xi^4}{2\left(1-\xi^2\right)\left(2-\xi^2\right)^2}\right], \quad \kappa_1 = \frac{\left(2-\xi^2\right)^2}{2\left(1-\xi^2\right)}. \tag{A2.18}$$

Let us substitute relation (A2.14) into the expression for Fermi's golden rule:

$$w_{\alpha'\alpha} = \frac{2\pi}{\hbar}\sum_{nn'}\sigma_n\left|\left\langle n\alpha\left|H_{\mathrm{int}}^{(1)}\right|n'\alpha'\right\rangle\right|^2 \delta\left(E_n - E_{n'} - \hbar\omega_{\alpha'\alpha}\right). \tag{A2.19}$$

The appearing therein squared matrix element of $H_{\text{int}}^{(1)}$ in the basis of phonon thermostat states $\{n_{\mathbf{q}\nu}\}$ is easily calculable using matrix elements of the second quantization operators:

$$\left|\left\langle \{n_{\mathbf{q}\nu}\},\alpha\left|H_{\text{int}}^{(1)}\right|\{n'_{\mathbf{q}\nu}\},\alpha'\right\rangle\right|^2 = \sum_{\mathbf{q}\nu}\left|\left\langle \alpha\left|\chi_{\mathbf{q}\nu}\right|\alpha'\right\rangle\right|^2\left[n_{\mathbf{q}\nu}\delta_{n'_{\mathbf{q}\nu},n_{\mathbf{q}\nu}-1} + \left(n_{\mathbf{q}\nu}+1\right)\delta_{n'_{\mathbf{q}\nu},n_{\mathbf{q}\nu}+1}\right]. \tag{A2.20}$$

Therefore, expression (A2.19) assumes the form:

$$w_{\alpha'\alpha} = \frac{2\pi}{\hbar}\sum_{\mathbf{q}\nu}\left|\left\langle \alpha\left|\chi_{\mathbf{q}\nu}\right|\alpha'\right\rangle\right|^2 \sum_{n_{\mathbf{q}\nu}n'_{\mathbf{q}\nu}} \frac{\exp\left[-\beta\hbar\omega_\nu(\mathbf{q})n_{\mathbf{q}\nu}\right]}{\sum_n \exp\left[-\beta\hbar\omega_\nu(\mathbf{q})n\right]}$$
$$\times\left[n_{\mathbf{q}\nu}\delta_{n'_{\mathbf{q}\nu},n_{\mathbf{q}\nu}-1} + (n_{q\nu}+1)\delta_{n'_{\mathbf{q}\nu},n_{\mathbf{q}\nu}+1}\right]\delta\left(\hbar\omega_\nu(\mathbf{q})(n_{\mathbf{q}\nu}-n'_{\mathbf{q}\nu}) - \hbar\omega_{\alpha'\alpha}\right). \tag{A2.21}$$

The summation over $n'_{\mathbf{q}\nu}$ retains only terms with $n'_{\mathbf{q}\nu} = n_{\mathbf{q}\nu} \pm 1$. Summing then over $n_{\mathbf{q}\nu}$, we arrive at the result:

$$\langle n\rangle = \frac{\sum_{n=0}^{\infty} n e^{-\beta\hbar\omega n}}{\sum_{n=0}^{\infty} e^{-\beta\hbar\omega n}} = \left(e^{\beta\hbar\omega} - 1\right)^{-1} \tag{A2.22}$$

As a consequence, expression (A2.21) turns into the sum of two quantities, namely, the probabilities for transitions per unit time accompanied by phonon absorption, $w_{\alpha'\leftarrow\alpha}^{(-)}$, and emission, $w_{\alpha'\leftarrow\alpha}^{(+)}$, which are expressed as:

$$w_{\alpha'\leftarrow\alpha}^{(\mp)} = \frac{2\pi}{\hbar^2}\left(\langle n_{\alpha'\alpha}\rangle + \frac{1}{2} \mp \frac{1}{2}\right)\sum_{q\nu}\left|\left\langle \alpha\left|\chi_{q\nu}\right|\alpha'\right\rangle\right|^2 \delta\left(\omega_{\alpha'\alpha} - \omega_\nu(\mathbf{q})\right) \tag{A2.23}$$

where

$$\langle n_{\alpha'\alpha}\rangle = \left[\exp\left(\hbar\left|\omega_{\alpha'\alpha}\right|/k_B T\right)-1\right]^{-1}, \quad \omega_{\alpha'\alpha} = \left(\varepsilon_{\alpha'}-\varepsilon_\alpha\right)/\hbar . \tag{A2.24}$$

We substitute relation (A2.15) or (A2.16) respectively for bulk or Rayleigh phonons into Eq. (A2.23) to obtain the same result:[164]

$$w^{(\mp)}_{\alpha'\leftarrow\alpha} = \kappa \frac{m_C^2\left|\omega_{\alpha'\alpha}\right|^5}{2\pi\hbar\rho\bar{c}^3}\left(\langle n_{\alpha'\alpha}\rangle + \frac{1}{2} \mp \frac{1}{2}\right)\left[\left(\mathbf{r}_\parallel\right)^2_{\alpha'\alpha} + \kappa_1\left(\mathbf{r}_z\right)^2_{\alpha'\alpha}\right]. \tag{A2.25}$$

For bulk phonons, $\kappa = \kappa_1 = 1$ and for Rayleigh phonons, the values κ and κ_1 are respectively specified by the equation $\kappa \equiv 3\pi/[8\lambda(2+\gamma^3)]$ and by Eq. (A2.18). The average sound velocity $\bar{c}$ was introduced in Eq. (A1.66).

The probability for a transition to the first excited state of rotational vibrations to occur per unit time is readily estimated by formula (A2.25) if the harmonic approximation with the matrix element $\left(\mathbf{r}_\parallel\right)^2_{10} = \hbar/2m_C\omega_{10}$ is invoked:

$$w \equiv w^{(-)}_{1\leftarrow 0} = \kappa \frac{m_C\omega_{10}^4}{4\pi\rho\bar{c}^3}\left[\exp\left(\hbar\omega_{10}/k_B T\right)-1\right]^{-1}. \tag{A2.26}$$

The comparison of formulae (A2.26) and (A2.4) proves to be informative. The temperature dependence of w is determined by the average number of phonons with the frequency equal to ω_{10}. Such a temperature-dependent cofactor typical of the Bose-Einstein statistics was invoked previously to estimate the reorientation frequency for light bulk solid-state defects. At $\hbar\omega_{10} \ll k_B T$, this cofactor provides a linear dependence $w(T)$ and at the inverse inequality changes to the Boltzmann factor $\exp(-\hbar\omega_{10}/k_B T)$ with the activation energy $\hbar\omega_{10}$ (instead of ΔU_φ in Eq. (A2.4)). The cofactor of the bracketed expression in Eq. (A2.26) depends on the parameters of a substrate and a diatomic group. Its value for the system OH/SiO_2 ($m_C = m_H = 1.67\cdot 10^{-24}$ g, $\omega_{10} = 3.77\cdot 10^{13}$ s^{-1} (200 cm^{-1}), $\rho = 2.2$ g/cm^3, $\bar{c} = 5.6\cdot 10^5$ cm/s, $\kappa = 1$) amounts to $6.9 \cdot 10^{11}$ s^{-1} (3.7 cm^{-1}). At $T = 300$ K, the probability (A2.26) is found to be $w = 4.4 \cdot 10^{11}$ s^{-1} (2.3 cm^{-1}), which is consistent with experimentally measured IR-absorption spectral line halfwidths for valence vibrations of hydroxyl groups on the silica surface. It should be noted that unlike Eq. (A2.4), relation (A2.26) is dependent on a specific mechanism of the subsystem-thermostat interaction.

A rigorous treatment of the IR-absorption spectral line broadening for valence vibrations of a reorienting group should include, in addition to reorientational

Brownian motion, other mechanisms, as for instance, a decay of a local vibration into substrate phonons (see Chapter 4) or inhomogeneous broadening caused by static shifts of oscillator frequencies in random electric fields of a disordered dipole environment. A temperature dependence of a broadening arising from these additional effects should be considerably weaker than the exponential dependence in Eq. (A2.26) or (A2.4). The total broadening is therefore expressible as

$$\Delta\omega_{1/2}(T) = \text{const} + w(T)\,. \tag{A2.27}$$

It behaves like $w(T)$ at high temperatures and becomes a constant at low temperatures, as illustrated in Fig. A2.2.

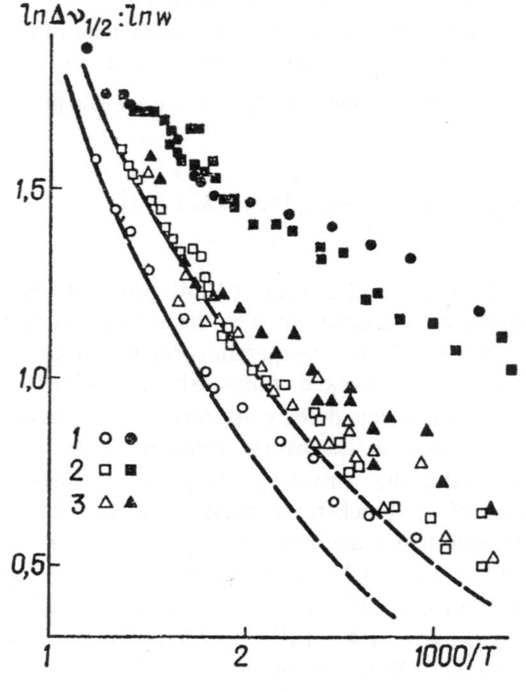

Fig. A2.2. Temperature dependences measured for the halfwidths $\Delta\nu_{1/2}$ of the IR absorption bands for valence vibrations of OH(D) groups on SiO_2 surface (filled markers) and recalculated for the halfwidths w of three components of Lorentzian lines (empty markers) for OH (1) and (OD) groups of high concentration (2), and for (OD) groups of low concentration (3).[202]

An additional factor implied by the summand "const" in Eq. (A2.27) is the multicomponent structure of the spectral line which results from the tunnel splitting of the vibrational levels approximating orientational states (see Fig. 4.4). Previously the temperature dependences $\Delta\nu_{1/2}(T)$ for OH(D) groups on the SiO_2 surface were measured and converted into $w(T)$ with regard for the relative shift of three components by 2.2 cm^{-1}.[202] The results are demonstrated in Fig. A2.2. The plotted curves correspond to the Bose-Einstein temperature factors in Eq. (A2.26) with the value $\omega_{10} = 3.77 \cdot 10^{13}$ s^{-1} ($\hbar\omega_{10} \approx 25$ meV). The inhomogeneous broadening exhibited in the low-temperature region was discussed in Refs. 61 and 121.

At sufficiently low temperatures ($k_B T << \hbar\omega_{10}$), the rate of transitions to the first excited state of rotational vibrations (see Eq. (A2.26)) becomes exponentially small, so that the transitions between the lowest tunnel-split energy levels predominate in the reorientation process. To be precise, the term "reorientation" keeps its original meaning no more, because the process under discussion is of purely quantum nature and represents the orientational delocalization of the atom C caused by the subbarrier tunneling in the interaction with the thermostat. The process of this kind was termed most accurately in the title of Ref. 214, namely, "tunneling relaxation in a phonon field". It has been shown in the above paper that in a certain low-temperature range, relaxation times are governed by two-phonon transitions. In other words, the transitions between tunnel-split levels are attributed to the interaction which is a quadratic function of atomic displacements in a phonon thermostat. An interaction appearing as expression (A2.13) is a linear function of the displacements **u** and gives rise to one-phonon transitions. It is therefore necessary to derive a more exact expression that involves, in addition to the terms of Eq. (A2.13), contributions quadratic in **u**.

Let $\mathbf{r}_j$ (j = 0, 1, 2, ...) and $\mathbf{r}_C$ respectively denote the positions of heavy atoms bound to each other and the position of the light atom C strongly bound to the atom with j = 0 and weakly bound to the nearest n atoms with j = 1,..., n. In Fig. A2.1, the atom with j = 0 is represented by the atom B and atoms with j = 1, ..., n are O_1,..., O_n (n = 3). The Hamiltonian function for the nuclei of the atoms concerned may be written in the approximation of pairwise interactions with the atom C:

$$H = \frac{m_C \dot{r}_C^2}{2} + \sum_j \frac{m_j \dot{r}_j^2}{2} + \sum_j U_{Cj}(\mathbf{r}_C - \mathbf{r}_j) + \tilde{U}(\mathbf{r}_0, \mathbf{r}_1, \ldots). \qquad \text{(A2.28)}$$

We now turn to the vectors of the inertia center, $\tilde{\mathbf{r}}_0 = (m_0\mathbf{r}_0 + m_C\mathbf{r}_C)/(m_0 + m_C)$, and of the relative nuclear positions $\mathbf{r} = \mathbf{r}_C - \mathbf{r}_0$. Then Eq. (A2.28) becomes:

$$H=\frac{m\dot{\mathbf{r}}^2}{2}+\frac{(m_0+m_C)\dot{\tilde{\mathbf{r}}}_0^2}{2}+\sum_{j\neq 0}\frac{m_j\dot{r}_j^2}{2}+U_{C0}(\mathbf{r})$$
$$+\sum_{j\neq 0}U_{Cj}\left(\tilde{\mathbf{r}}_0-\mathbf{r}_j+\frac{m_0}{m_0+m_C}\mathbf{r}\right)+\tilde{U}\left(\tilde{\mathbf{r}}_0-\frac{m_C}{m_0+m_C}\mathbf{r},\mathbf{r}_1,\ldots\right). \tag{A2.29}$$

Expanding the sum of interactions U_{Cj} in $\mathbf{r}$ (see Eq. (1.56)) yielded, with the nearest n atoms positioned symmetrically, the hindered rotation potential

$$U(\varphi)=\frac{1}{2}\Delta U_\varphi[1-\cos n(\varphi-\varphi_0)]. \tag{A2.30}$$

Now we should make allowance for possible atomic displacements $\mathbf{u}_j$ from equilibrium positions $\mathbf{r}_j^{(0)}$ and perform an expansion both in $\mathbf{u}_j$ and $\mathbf{r}$ together. In so doing, $\mathbf{u}_0$ is considered to mean the displacement of the inertia center from the equilibrium position $\tilde{\mathbf{r}}_0^{(0)}$:

$$\tilde{U}\left(\tilde{\mathbf{r}}_0-\frac{m_C}{m_0+m_C}\mathbf{r},\mathbf{r}_1,\ldots\right)=\frac{1}{2}\left(\frac{m_C}{m_0+m_C}\right)^2\Phi_{00}^{\alpha\beta}r^\alpha r^\beta$$
$$-\frac{m_C}{m_0+m_C}r^\alpha\sum_{j\neq 0}\Phi_{0j}^{\alpha\beta}u_j^\beta+\frac{1}{2}\sum_{jj'}\Phi_{jj'}^{\alpha\beta}u_j^\alpha u_{j'}^\beta+\ldots, \tag{A2.31}$$

$$\sum_{j\neq 0}U_{Cj}\left(\tilde{\mathbf{r}}_0-\mathbf{r}_j\frac{m_0}{m_0+m_C}\mathbf{r}\right)=\sum_{j\neq 0}U_{Cj}(\Delta\mathbf{r}_j)$$
$$+\sum_{j\neq 0}\frac{\partial U_{Cj}(\Delta r_j)}{\partial\Delta r_j^\alpha}\left(u_0^\alpha-u_j^\alpha\frac{m_0}{m_0+m_C}r^\alpha\right)$$
$$+\frac{1}{2!}\sum_{j\neq 0}\frac{\partial^2U_{Cj}(\Delta r_j)}{\partial\Delta r_j^\alpha\partial\Delta r_j^\beta}\left(u_0^\alpha-u_j^\alpha\frac{m_0}{m_0+m_C}r^\alpha\right)\left(u_0^\beta-u_j^\beta\frac{m_0}{m_0+m_C}r^\beta\right)+\ldots, \tag{A2.32}$$
$$\Delta\mathbf{r}_j\equiv\tilde{\mathbf{r}}_0^{(0)}-\mathbf{r}_j^{(0)}.$$

Here $\Phi_{jj'}^{\alpha\beta}$ are the force constants. The first term in Eq. (A2.31) does not contain $\mathbf{u}_j$ and can be included in the potential $U_{C0}(\mathbf{r})$. The second term gives, with regard to the equation of motion for $\mathbf{u}_0$,

$$(m_0 + m_C)\ddot{u}_0^\alpha = -\sum_{j\neq 0} \Phi_{0j}^{\alpha\beta} u_j^\beta \,, \tag{A2.33}$$

the potential energy for the d'Alembert inertia force, $m_C\mathbf{r}\cdot\ddot{\mathbf{u}}_0$, as in Eq. (A2.13). The third term appears as a standard form of the potential energy for atomic displacements in the harmonic approximation. The subsequent expansion terms involve products of the different powers of the components r^α and u_j^α with the largest contribution to two-phonon interactions coming from the terms proportional to $r^\alpha u_j^\beta u_j^\gamma$. At $m_C << m_0$, they are small in the parameter m_C/m_0.

In the expansion (A2.32), the first term is merely a constant, while the second one renormalizes equilibrium atomic positions but gives no contribution to the interaction of the atom C with a thermostat (provided a symmetric disposition of atoms, the term linear in $\mathbf{r}$ vanishes). The third term contains small corrections to $U_{C0}(\mathbf{r})$ and to force constants $\Phi_{ij}^{\alpha\beta}$, and the contribution to a single-phonon interaction proportional to $r^\alpha\left(u_0^\beta - u_j^\beta\right)$. It is clear that in the continual approximation the quantities of such a structure will account for the deformational interaction, $\Gamma_{\alpha\beta}(\mathbf{r})u_{\alpha\beta}$ in Eq. (A2.13). The subsequent expansion terms in Eq. (A2.32) will contribute also to two-phonon interactions proportional to $r^\alpha(u_0^\beta - u_j^\beta)(u_0^\gamma - u_j^\gamma)$.

One-phonon interactions are described in general by formula (A2.14). An analogous expression for two-phonon interactions appears as:

$$H_{\rm int}^{(2)} = \frac{1}{2}\sum_{\mathbf{q}\nu,\mathbf{q}'\nu'} \frac{B(\mathbf{q}\nu,\mathbf{q}'\nu)}{\sqrt{\omega_\nu(\mathbf{q})\omega_{\nu'}(\mathbf{q}')}}\left(b_{\mathbf{q}\nu} + b_{-\mathbf{q}\nu}^+\right)\left(b_{\mathbf{q}'\nu'} + b_{-\mathbf{q}'\nu'}^+\right), \tag{A2.34}$$

where B depends on the radius vector $\mathbf{r}$. Substitute Eq. (A2.34) into formula (A2.19) (with $H_{\rm int}^{(2)}$ instead of $H_{\rm int}^{(1)}$ and $\alpha' = 2$, $\alpha = 1$) that specifies rates of transitions between the two tunnel-split levels separated by the energy gap $\Delta\varepsilon = \hbar\omega_{21}$. In the case $\Delta\varepsilon << k_BT$, characteristic phonon energies E_n are much larger than $\Delta\varepsilon$ and hence the term $\hbar\omega_{21}$ in the argument of δ-function (see Eq. (A2.19)) can be omitted.

By calculating matrix elements of phonon operators in (A2.34) and summing over the occupation numbers $n_{\mathbf{q}\nu}$ (as performed in Eqs. (A2.20) and (A2.21)), we get:

$$w_{21}^{(2)} = \frac{\pi}{\hbar^2} \sum_{\mathbf{q}\nu, \mathbf{q}'\nu'} \frac{\left|\langle 1 | B(\mathbf{q}\nu, \mathbf{q}'\nu') | 2 \rangle\right|^2}{\omega_\nu(\mathbf{q})\omega_{\nu'}(\mathbf{q}')} \langle n_{\mathbf{q}\nu}\rangle \left(\langle n_{\mathbf{q}'\nu'}\rangle + 1\right) \delta(\omega_\nu(\mathbf{q}) - \omega_{\nu'}(\mathbf{q}')). \quad \text{(A2.35)}$$

To estimate the above expression, consider a particular case of the symmetric two-well potential in which the two-phonon interaction $H_{\text{int}}^{(2)}$ can approximately be derived from expansion (A2.32):

$$H_{\text{int}}^{(2)} \approx \frac{\Delta U_\varphi}{2r\Delta r_1} \cos\varphi \left(u_1^x - u_2^x\right)\left(2u_2^x - u_1^x - u_2^x\right). \quad \text{(A2.36)}$$

(The axis x goes through atoms with j = 1, 2 and φ is the angle between this axis and the vector **r**). We use Eq. (A1.43) for a monoatomic crystal to transform Eq. (A2.36) to (A2.34) with the quantity $B(\mathbf{q}\nu, \mathbf{q}'\nu')$ appearing as:

$$\begin{aligned} B(\mathbf{q}\nu, \mathbf{q}'\nu) &= \frac{\hbar \Delta U_\varphi e_{x\nu}(\mathbf{q}) e_{x\nu'}(\mathbf{q}')}{2m_0 r \Delta r_1 N} \cos\varphi \left(e^{-i\mathbf{q}\Delta r_2} - e^{i\mathbf{q}\Delta r_2}\right) \\ &\times \left(2 - e^{-i\mathbf{q}'\Delta r_2} - e^{i\mathbf{q}'\Delta r_2}\right) \sim \frac{i\hbar \Delta U_\varphi \Delta r_1^2}{3m_0 r N} \left|q_x q_x'^2\right| \cos\varphi. \end{aligned} \quad \text{(A2.37)}$$

Substituting Eq. (A2.37) into (A2.35) and taking into consideration that $\langle 1|\cos\varphi|2\rangle \approx 1$, we obtain, in the Debye approximation at $\Delta\varepsilon / k_B << T << T_D$:

$$w_{21}^{(2)} \approx \frac{2^5 \cdot 35 \cdot \varsigma(8) \Delta U_\varphi^2 \Delta r_1^4}{\hbar^9 \rho^2 r^2 \bar{c}^{12}} T^9. \quad \text{(A2.38)}$$

The same temperature dependence for symmetric wells was reported previously (for asymmetric wells, $w_{21}^{(2)} \propto T^7$).[214]

Compare Eq. (A2.38) with the contribution from one-phonon transitions for which we can apply formula (A2.25) with $\left(r_\parallel\right)_{12}^2 = r^2$:

$$w_{21}^{(1)} \approx \frac{m_C^2 r^2}{2\pi\hbar^2 \rho \bar{c}^3} \left(\frac{\Delta\varepsilon}{\hbar} \right)^4 T. \qquad \text{(A2.39)}$$

With the parameters $\Delta\varepsilon = 0.28$ meV, $\Delta U_\varphi = 50$ meV, $\Delta r_l = 3r = 3$Å, $\bar{c} = 5.6 \cdot 10^5$ cm/s, $\rho = 2.2$ g/cm^3, expressions (A2.38) and (A2.39) prove to be equal to 10^4 c^{-1} (10^{-7} cm^{-1}) at $T = T^* \sim 30$ K, while expressions (A2.26) and (A2.38) become equal at $T = T^{**} \sim 170$ K.

Thus, at temperatures $T < T^*$, the one-phonon relaxation described by Eq. (A2.39) is dominant, in the temperature range $T^* < T < T^{**}$, the two-phonon relaxation (see Eq. (A2.38)) prevails, and at $T > T^{**}$, rotational vibrations are excited and the relaxation is of thermoactivational nature as seen from Eq. (A2.26). It is noteworthy that the $w_{21}^{(2)}$ (and especially $w_{21}^{(1)}$) is several orders of magnitude smaller than ω_{21} for the atom C with a mass of a hydrogen atom. The dependence of $\omega_{21} = \Delta\varepsilon/\hbar$ on m_C is given by the formula

$$\Delta\varepsilon \approx \kappa(n) \frac{n^2\hbar^2}{\pi I} e^{1/2} 2^{5/2} p^{3/2} e^{-4p}, \quad \kappa(n) = \begin{cases} 1, & n - \text{even} \\ 3/4 - n = 3 \end{cases} \qquad \text{(A2.40)}$$

which implies that the value ω_{21} for a heavy atom C may be several orders smaller and only then it may be equal to $w_{21}^{(2)}$. The relaxation times for a two-level subsystem are found which at $w_{21} << \omega_{21}$ are equal to $2w_{21}$ for diagonal elements of the density matrix (in the basis of eigenfunctions of the subsystem) and to w_{21} for nondiagonal elements that decay oscillating with the frequency ω_{21}.[61] As was shown before,[214] at $w_{21} >> \omega_{21}$, the subsystem is characterized by quite different relaxation times. For surface atomic groups with a rotational degree of freedom, the inverse inequality, $w_{21} << \omega_{21}$, holds and the relaxation times differ only by a factor of two.

In this Appendix, we attempted to elucidate the basic features of the transition of a particle from one potential well to another in various special cases: at high reduced barriers p when a classical description is applicable, for $p \sim 1$ when the probability for a rotational vibration to occur can be regarded as an estimate of the reorientation frequency for not too low temperatures, and, finally, for the low-temperature situation when subbarrier tunneling relaxation becomes dominant. However, the quantitative description of the processes considered cannot be taken as satisfactory, since it is rather fragmentary and, in addition, no general expression has been derived for an average reorientation frequency which would reduce to Eqs. (A2.4), (A2.26) or to (A2.38), (A2.39) in the above-listed special cases. As of now, an adequate approach has been developed which allows this quantity to be calculated in a wide range of parameter values.[215] The subbarrier tunneling of a particle is described by the equations of motion dependent on imaginary time. Solutions of these equations specify possible trajectories for a particle interacting with a thermostat. Then the technique of Feynman integrals over trajectories is invoked. Extremum (action-minimizing) trajectories of a certain kind which correspond to a transition of a particle from one potential well into another are called instantons.[216] In quantum chronodynamics, instantons are solutions of the field equations with a nontrivial topology and allow for gluon field fluctuations which cannot be taken into consideration by the perturbation theory.

The points discussed in this appendix are of direct relevance to a description of the chemical reactions that involve surmounting the potential barrier. The barrier is surmounted due to thermal activation at high temperatures and due to tunneling at low temperatures, the latter fact accounting for an occurrence of low-temperature chemical reactions.[216,217]

Appendix 3

The spectral function of local vibration (low-temperature approximation involving collectivized high-frequency and low-frequency modes)

In the Heitler-London approximation, with allowance made only for biquadratic anharmonic coupling between collectivized high-frequency and low-frequency modes of a lattice of adsorbed molecules (admolecular lattice), the total Hamiltonian (4.3.1) can be written as a sum of harmonic and anharmonic contributions:

$$\hat{H}_{tot} = \hat{H} + \hat{H}_A, \qquad \hat{H} = \sum_{\mathbf{K}} \hat{H}_{\mathbf{K}}, \tag{A3.1}$$

$$\hat{H}_{\mathbf{K}} = \hbar\Omega_{\mathbf{K}} a_{\mathbf{K}}^{+} a_{\mathbf{K}} + \hbar\omega_{\mathbf{K}} b_{\mathbf{K}}^{+} b_{\mathbf{K}} + \hbar\sum_{\nu} \omega_{\mathbf{K},\nu} \beta_{\mathbf{K},\nu}^{+} \beta_{\mathbf{K},\nu} + \hbar\sum_{\nu} (\chi_{\mathbf{K},\nu} b_{\mathbf{K}}^{+} \beta_{\mathbf{K},\nu} + h.c.), \tag{A3.2}$$

$$\hat{H}_A = \frac{\hbar\gamma}{N} \sum_{\mathbf{K}_1\mathbf{K}_2\mathbf{K}_3} a_{\mathbf{K}_1}^{+} a_{\mathbf{K}_2} b_{\mathbf{K}_3}^{+} b_{\mathbf{K}_1-\mathbf{K}_2+\mathbf{K}_3}, \tag{A3.3}$$

where the terms in the expression (A3.2) respectively represent high-frequency and low-frequency vibrations of the admolecular lattice, crystal phonons, and the harmonic coupling between phonons and low-frequency modes. In the above relation, quantum states of phonons of the admolecular lattice are characterized by the surface-parallel wave vector **K**, whereas the quantum numbers of substrate phonons are indicated by the couple of indices **K** and ν. The latter accounts for the polarization of a quasi-particle and its motion in the surface-normal direction; it implicitly reflects both the atomic arrangement in the crystal unit cell and the position of the admolecular lattice, as a whole, relative to the crystal surface. The utility of introduction of the complex index ν was substantiated previously.[138]

As a rule, the density of states for molecular lattice vibrations is negligible as compared to that for crystal phonons. Therefore, the **K**-mode of a molecular lattice is coupled with the crystal phonons specified by the same wave vector **K**. Besides, the low-frequency collective mode $\omega_{\mathbf{K}}$ of adsorbed molecules can be considered as a resonance vibration with the renormalized frequency $\tilde{\omega}_{\mathbf{K}}$ and the inverse lifetime $\eta_{\mathbf{K}}(\tilde{\omega}_{\mathbf{K}})$.[138,146]

The response of the system concerned to an external electromagnetic field is conveniently described in terms of double-time Green's function (GF) which can be introduced in a variety of representations.[144,218-221] In what follows we will involve the representation in Matsubara's frequency space[218] which is accepted in the theory of anharmonic crystals[197] and provides a number of exact solutions in the case of a single adsorbed molecule.[150,152] In this approach, the spectral line shape for high-frequency vibrations can be determined as follows:[184]

$$L(\omega) = \frac{\hbar\beta}{\pi} \operatorname{Im} G(a_0 a_0^+, \omega + i0), \; \beta = (k_B T)^{-1}, \tag{A3.4}$$

$$G(a_0 a_0^+, \omega) = \frac{1}{\beta} \int_0^\beta G(a_0 a_0^+, \tau) e^{\hbar\omega\tau} d\tau , \tag{A3.5}$$

$$G(a_0 a_0^+, \tau) = \frac{\left\langle \hat{\mathcal{T}} \tilde{a}_0(\tau) \tilde{a}_0^+(0) \hat{S}(\beta) \right\rangle_0}{\left\langle \hat{S}(\beta) \right\rangle_0}, \tag{A3.6}$$

where

$$\hat{S}(\beta) \equiv \hat{S}(\beta, 0) , \quad \hat{S}(\tau, \tau_0) = \hat{\mathcal{T}} \exp\left\{ - \int_{\tau_0}^{\tau} d\tau' \tilde{H}_A(\tau') \right\}, \tag{A3.7}$$

$$\tilde{A}(\tau) = e^{\hat{H}\tau} \hat{A} e^{-\hat{H}\tau} , \qquad \langle \ldots \rangle_0 = \frac{Sp\left\{ e^{-\beta \hat{H}} \ldots \right\}}{Sp\left\{ e^{-\beta \hat{H}} \right\}} \tag{A3.8}$$

and $\hat{A}$ is an arbitrary operator function.

Some of lowest-order diagrams for the temperature GF (A3.6) are shown in Fig. A3.1. The dashed and the solid lines represent the GFs of high-frequency and low-frequency vibrations of a planar lattice in the harmonic approximation:

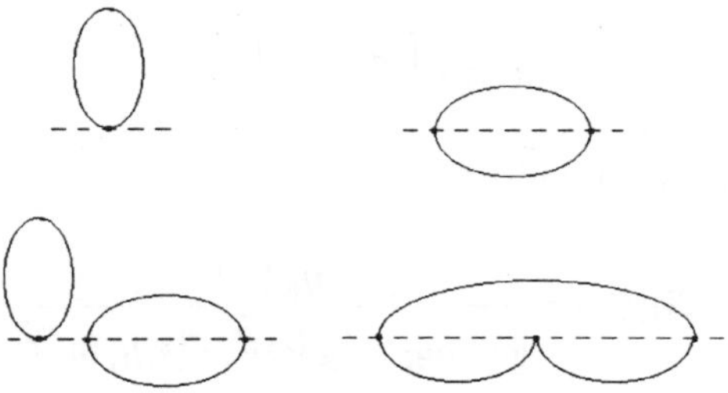

a

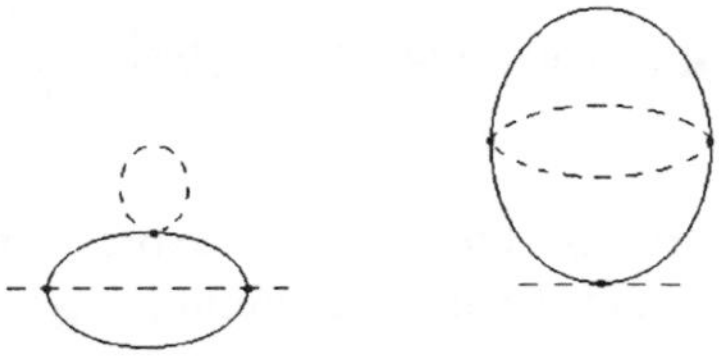

b

Fig. A3.1. Some lowest-order diagrams for the temperature GF (A3.6). The dashed and solid lines correspond to the GF for high-frequency and resonance low-frequency vibrations of a molecular planar lattice in the harmonic approximation (see Eq. (A3.9) and (A3.10)). Each vertex is associated with the factor $-\gamma/N$, the integration and summation being performed over each vertex coordinates τ_i from 0 to β, and over all internal wave vectors **K**. At $\beta\hbar\Omega_{\mathbf{K}} >> 1$, the main contribution is provided by a-type diagrams.[184]

$$G_{\mathbf{K}}^{(0)}(\tau) = \left\langle \hat{\mathcal{T}}\, a_{\mathbf{K}}(\tau) a_{\mathbf{K}}^{+}(0) \right\rangle_0 = \{[n(\Omega_{\mathbf{K}})+1]\theta(\tau) + n(\Omega_{\mathbf{K}})\theta(-\tau)\} e^{-\hbar\Omega_{\mathbf{K}}\tau}, \qquad \text{(A3.9)}$$

$$g_{\mathbf{K}}^{(0)}(\tau) = \left\langle \hat{\mathcal{T}}\, \tilde{b}_{\mathbf{K}}(\tau) \tilde{b}_{\mathbf{K}}^{+}(0) \right\rangle_0 = \int_{-\infty}^{\infty} d\omega \Re_{\mathbf{K}}(\omega) \{[n(\omega)+1]\theta(\tau) + n(\omega)\theta(-\tau)\} e^{-\hbar\omega\tau},$$

(A3.10)

where

$$n(\omega) = \left[e^{\beta \hbar \omega} - 1\right]^{-1} \tag{A3.11}$$

is the factor of Bose-Einstein statistics and

$$\Re_{\mathbf{K}}(\omega) = \frac{\widetilde{\eta}_{\mathbf{K}}(\omega)}{\left[\omega - \omega_{\mathbf{K}} - \widetilde{\mathrm{P}}_{\mathbf{K}}(\omega)\right]^2 + \left[\pi \widetilde{\eta}_{\mathbf{K}}(\omega)\right]^2} \tag{A3.12}$$

is the resonance function of low-frequency vibrations,[138,140]

$$\widetilde{\eta}_{\mathbf{K}}(\omega) = \sum_{\nu} \left|\chi_{\mathbf{K},\nu}\right|^2 \delta(\omega - \omega_{\mathbf{K},\nu}), \quad \widetilde{P}_{\mathbf{K}}(\omega) = \int_{-\infty}^{\infty} \frac{\widetilde{\eta}_{\mathbf{K}}(\widetilde{\omega}) d\widetilde{\omega}}{\omega - \widetilde{\omega}}. \tag{A3.13}$$

The real and imaginary parts of the pole in expression (A3.12) define the renormalized frequency $\widetilde{\omega}_{\mathbf{K}}$ and the inverse lifetime $\eta_{\mathbf{K}}(\widetilde{\omega}_{\mathbf{K}}) = 2\pi\, \widetilde{\eta}_{\mathbf{K}}(\widetilde{\omega}_{\mathbf{K}})$ of the resonance vibration.

Let us take advantage of the inequality $\beta \hbar\, \Omega_{\mathbf{K}} >> 1$ permitting neglect of the terms of the order $\exp(-\beta\hbar\Omega_{\mathbf{K}})$. Then the trace taken over high-frequency and low-frequency modes, Sp{...}, is reduced to that for low-frequency modes, with all high-frequency vibrations considered only for the ground state. The resulting temperature GF (A3.6) takes the form:

$$G(a_0 a_0^+, \tau) = \left\langle \widetilde{a}_0(\tau) \hat{S}(\tau) \widetilde{a}_0^+(0) \right\rangle_0, \quad \beta > \tau > 0\,. \tag{A3.14}$$

This expression refers to diagrams without closed high-frequency loops (Fig. A3.1 a). Thus, provided the inequality $\beta\hbar\Omega_{\mathbf{K}} >> 1$ is valid, the GF defined above can be written in the form involving no high-frequency mode operators:

$$G(a_0 a_0^+, \tau) = e^{-\hbar\Omega_0 \tau} \left\langle \hat{s}_{00}(\tau) \right\rangle_0, \tag{A3.15}$$

$$\hat{\mathbf{s}}_{\mathbf{KK}'}(\tau)=\sum_{n=0}^{\infty}(-\hbar\gamma)^n\int_0^{\tau}d\tau_1\int_0^{\tau_1}d\tau_2...\int_0^{\tau_{n-1}}d\tau_n\sum_{\mathbf{K}_1...\mathbf{K}_{n-1}}\widetilde{B}_{\mathbf{KK}_1}(\tau_1)\widetilde{B}_{\mathbf{K}_1\mathbf{K}_2}(\tau_2)...\widetilde{B}_{\mathbf{K}_{n-1}\mathbf{K}'}(\tau_n)$$
$$=\left(\hat{\mathcal{T}}\exp\left\{-\hbar\gamma\int_0^{\tau}\hat{B}(\tau_1)d\tau_1\right\}\right)_{\mathbf{KK}'},\qquad\text{(A3.16)}$$

$$\widetilde{B}_{\mathbf{KK}'}(\tau)\equiv\left(\hat{B}(\tau)\right)_{\mathbf{KK}'}=\frac{1}{N}\exp[\hbar(\Omega_{\mathbf{K}}-\Omega_{\mathbf{K}'})\tau]\sum_{\mathbf{K}_1}\widetilde{b}^{+}_{\mathbf{K}_1-\mathbf{K}}(\tau)\widetilde{b}_{\mathbf{K}_1-\mathbf{K}'}(\tau)\,.\qquad\text{(A3.17)}$$

Representing average value (A3.15) as

$$\langle\hat{s}_{00}(\tau)\rangle_0=e^{F(\tau)}\qquad\text{(A3.18)}$$

and differentiating the left and the right sides of expression (A3.18) with respect to τ, we arrive at:

$$\frac{dF(\tau)}{d\tau}=-\hbar\gamma\sum_{\mathbf{K}}\frac{\left\langle\widetilde{B}_{0\mathbf{K}}(\tau)\hat{s}_{\mathbf{K}0}(\tau)\right\rangle_0}{\langle\hat{s}_{00}(\tau)\rangle_0}\,.\qquad\text{(A3.19)}$$

Eq. (A3.19) corresponds to compact diagrams in Fig. A3.1a. The expression obtained allows the line shape to be determined in the approximation of the small anharmonic coefficient γ, as well as in the low-temperature limit.

Importantly, the three-time GF

$$g_{\mathbf{K},\mathbf{K}'}(\tau,\tau',\tau'')=\frac{\left\langle\hat{\mathcal{T}}\,\widetilde{b}_{\mathbf{K}}(\tau)\widetilde{b}^{+}_{\mathbf{K}'}(\tau')\hat{s}_{\mathbf{K}'-\mathbf{K},0}(\tau'')\right\rangle_0}{\langle\hat{s}_{00}(\tau'')\rangle_0}\qquad\text{(A3.20)}$$

satisfies the Dyson-like equation with the assumption of non-interacting high-frequency collective vibrations ($\Omega_{\mathbf{K}}=\Omega_0$):

$$g_{\mathbf{K},\mathbf{K}'}(\tau,\tau',\tau'')=g_{\mathbf{K}}^{(0)}(\tau-\tau')\delta_{\mathbf{K},\mathbf{K}'}-\frac{\hbar\gamma}{N}\sum_{\mathbf{K}_1}\int_0^{\tau''}d\tau_1 g_{\mathbf{K}}^{(0)}(\tau-\tau_1)g_{\mathbf{K}_1,\mathbf{K}'}(\tau_1,\tau',\tau'') \tag{A3.21}$$

(the resonance GF $g_{\mathbf{K}}^{(0)}(\tau)$ in the harmonic approximation is determined by the expression (A3.10)). Equation (A3.21) takes into account intermolecular interactions of low-frequency resonance modes and thus generalizes the corresponding equations derived before.[150,152]

To invoke the perturbation theory for a small anharmonic coupling coefficient, we use the Wick theorem for the coupling of the creation and annihilation operators of low-frequency modes in expression (A3.19). Retaining the terms of the orders γ and γ^2, we are led to the following expressions for the shift $\Delta\Omega$ and the width 2Γ of the high-frequency vibration spectral line:[184]

$$\Delta\Omega=\frac{\gamma}{N}\sum_{\mathbf{K}}\int_{-\infty}^{\infty}d\omega n(\omega)\Re_{\mathbf{K}}(\omega), \tag{A3.22}$$

$$2\Gamma=\frac{2\pi\gamma^2}{N^2}\sum_{\mathbf{K},\mathbf{K}'}\int_{-\infty}^{\infty}d\omega\, n(\omega)[n(\omega+\Omega_0-\Omega_{\mathbf{K}})+1]\Re_{\mathbf{K}+\mathbf{K}'}(\omega)\Re_{\mathbf{K}'}(\omega+\Omega_0-\Omega_{\mathbf{K}}), \tag{A3.23}$$

where $n(\omega)$ and $\Re_{\mathbf{K}}(\omega)$ are respectively determined by relations (A3.11) and (A3.12). It should be noted that expressions (A3.22) and (A3.23) describe a single Lorentz-like spectral line of local vibrations, with its position and width dictated by dispersion laws and lifetimes for resonance low-frequency modes.

The analytical description of high-frequency line shapes becomes possible in the low-temperature limit, i.e., at $n(\omega_{\mathbf{K}}) \approx \exp\{-\beta\hbar\omega_{\mathbf{K}}\}<<1$, which represents an experimentally important case. In this situation, the Wick coupling for the operators of low-frequency modes in expression (A3.19) involves only the terms in which the annihilation operator is to the left of the creation operator in all but one operator pair. Then Eq. (A3.19) can be written as:

$$\frac{dF(\tau)}{d\tau} = -\frac{\hbar\gamma}{N}\sum_{\mathbf{K}}\int_{-\infty}^{\infty} d\omega \Re_{\mathbf{K}}(\omega) n(\omega) \sum_{n=0}^{\infty}(-\hbar\gamma)^n \int_0^{\tau} d\tau_1 \varphi_{\mathbf{K}}(\omega, \tau - \tau_1)$$
$$\times \int_0^{\tau_1} d\tau_2 \varphi_{\mathbf{K}}(\omega, \tau_1 - \tau_2) \ldots \int_0^{\tau_{n-1}} d\tau_n \varphi_{\mathbf{K}}(\omega, \tau_{n-1} - \tau_n), \tag{A3.24}$$

where

$$\varphi_{\mathbf{K}}(\omega, \tau) = \frac{1}{N}\sum_{\mathbf{K}'}\int_{-\infty}^{\infty} d\omega' \Re_{\mathbf{K}'}(\omega') \exp\left[\hbar\left(\Omega_0 - \Omega_{\mathbf{K}-\mathbf{K}'} + \omega - \omega'\right)\tau\right]$$
$$\approx \frac{1}{N}\sum_{\mathbf{K}'} \exp\left\{\hbar\left[\Omega_0 - \Omega_{\mathbf{K}-\mathbf{K}'} + \omega - \omega_{\mathbf{K}'} + i\pi\tilde{\eta}(\omega_{\mathbf{K}'})\right]\tau\right\}. \tag{A3.25}$$

On integrating Eq. (A3.24) over the variables τ_1, τ_2, ..., τ_n, we obtain:

$$\frac{dF(\tau)}{d\tau} = -\frac{\hbar\gamma}{2\pi N}\sum_{\mathbf{K}}\int_{-\infty}^{\infty} d\omega \Re_{\mathbf{K}}(\omega) n(\omega) \int_{-\infty}^{\infty} dy \frac{e^{\tau(\lambda + iy)}}{\lambda + iy}$$
$$\times \left[1 + \hbar\gamma \int_0^{\infty} \varphi_{\mathbf{K}}(\omega, x) e^{-x(\lambda + iy)} dx\right]^{-1}. \tag{A3.26}$$

Then expression (A3.26) in the Markov approximation ($\hbar\tilde{\eta}_{\mathbf{K}}(\omega)\tau$, $\hbar\gamma\tau >> 1$) can be reduced as follows:[184]

$$\frac{dF(\tau)}{d\tau} = -\frac{\hbar\gamma}{N}\sum_{\mathbf{K}} \frac{n(\omega_{\mathbf{K}})}{1 - (\gamma/N)\sum_{\mathbf{K}'}\left[\alpha_{\mathbf{KK}'} + i\pi\left(\tilde{\eta}_{\mathbf{K}}(\omega_{\mathbf{K}}) + \tilde{\eta}_{\mathbf{K}'}(\omega_{\mathbf{K}'})\right)\right]^{-1}}, \tag{A3.27}$$

where

$$\alpha_{\mathbf{KK}'} = \Omega_0 - \Omega_{\mathbf{K}-\mathbf{K}'} + \omega_{\mathbf{K}} - \omega_{\mathbf{K}'}. \tag{A3.28}$$

As seen, the spectral line of high-frequency local vibrations is of the Lorentz-like shape:

$$L(\omega) = -\frac{1}{\pi}\operatorname{Im}\frac{1}{\omega - \Omega_0 - W}, \qquad W = -\frac{dF(\tau)}{\hbar d\tau}, \tag{A3.29}$$

with the shift $\Delta\Omega = \operatorname{Re} W$ and the width $2\Gamma = -2\operatorname{Im} W$.

Bibliography*

1. K. Burke, D. C. Langreth, M. Persson, and Z.-Y. Zhang, Phys. Rev. B **47**, 15869 (1993). {6, 105, 111, 112}
2. W. Steele, Chem. Rev. **93**, 2355 (1993). {6, 51}
3. D. Marx, H. Wiechert, Adv. Chem. Phys. **95**, 213 (1996). {6, 51}
4. A. Patrykiejew, S. Sokolowski, K. Binder, Surface Science Reports **37**, 207 (2000). {6, 51}
5. R. D. Diehl, M. F. Toney, and S. C. Fain, Jr., Phys. Rev. Lett. **48**, 177 (1982). {6, 8, 51}
6. R. D. Diehl and S. C. Fain, Jr., Phys. Rev. B **26**, 4785 (1982). {6, 8}
7. R. Wang, S.-K. Wang, H. Taub, J. C. Newton and H. Shechter, Phys. Rev. B **35**, 5841 (1987). {6, 8}
8. S.-K. Wang, J. C. Newton, R. Wang, H. Taub, J. R. Dennison and H. Shechter, Phys. Rev. B **39**, 10331 (1989). {6, 8}
9. M. F. Toney and S. C. Fain, Jr., Phys. Rev. B **36**, 1248 (1987). {6, 8}
10. K. Morishige, Mol. Phys. **78**, 1203 (1993). {6, 8}
11. A. Terlain, Y. Larher, F. Angerand, G. Parette, H. Lauter, I.C. Bassignana, Mol. Phys. **58**, 799 (1986). {6, 8}
12. Y. P. Joshi, D. J. Tildesley, J. S. Ayres, R. K. Thomas, Mol. Phys. **65**, 991 (1988). {6, 8}
13. O. Berg and G. E. Ewing, Surf. Sci. **220**, 207 (1989). {6, 8, 9, 71}
14. O. Berg, R. Disselkamp and G. Ewing, Surf. Sci. **277**, 8 (1992). {6, 8, 9, 71}
15. J. Heidberg, E. Kampshoff, O. Schönekäs, H. Stein and H. Weiss, Ber. Bunsen-Ges. Phys. Chem. **94**, 112, 118, 127 (1990). {6, 8, 9, 71}
16. J. Heidberg, E. Kampshoff, R. Kühnemuth and O. Schönekäs,. Surf. Sci. **272**, 306 (1992). {6, 8, 9, 71}
17. J. Heidberg, E. Kampshoff, R. Kühnemuth and O. Schönekäs,. Surf. Sci. **269/270**, 120 (1992). {6, 8, 9, 71}
18. J. Schimmelpfennig, S. Folsch, and M. Henzler, Verh. Deutsch Phys. Ges. **4**, 192 (1990). {6, 8, 71}
19. G. Lange, J. P. Toennies, R. Vollmer, and H. Weiss, J. Chem. Phys. **98**, 10096 (1993). {6, 8, 71}
20. J. Heidberg, E. Kampshoff, R. Kuhnemuth, O. Schonekas, M. Suhren, J. Electron. Spect. Related Phenom., **54/55**, 945 (1990). {9, 71}
21. W. Chen, W.L. Schaich, Surf. Sci. **220**, L733 (1989). {9, 71}

* Pages on which the references appear in the book are indicated in braces.

22. G. Lange, D. Schmicker, J. P. Toennies, R. Vollmer, and H. Weiss, J. Chem. Phys. **103**, 2308 (1995). {9, 71}
23. J. Schimmelpfennig, S. Folsch, M. Henzler, Surf. Sci. **250**, 198 (1991). {9, 71}
24. A. Vigiani, G. Cardini, V. Schettino, J. Chem. Phys. **106**, 5693 (1997). {9, 71}
25. S. Picaud, S. Briquez, A. Lakhlifi, C. Girardet, J. Chem. Phys. **102**, 7229 (1995). {9, 71}
26. C. Girardet, S. Picaud, P.N.M. Hoang, Europhys. Lett. **25**, 131 (1994). {9, 71}
27. J. Heidberg, K. Stahmer, H. Stein, H. Weiss, M. Folman, Z. Phys. Chem. **755**, 223 (1987). {9}
28. J. Heidberg, E. Kampshoff, R. Kuhnemuth, M. Suhren, and H. Weiss, Surf. Sci. **269/270**, 128 (1992). {9, 29, 42, 67, 71}
29. R. Disselkamp, H.-C. Chang, G. E. Ewing, Surf. Sci. **240**, 193 (1990). {9}
30. D. J. Dai, G. E. Ewing, J. Electron Spectrosc. Relat. Phenom. **64/65**, 101 (1993). {9}
31. S. Picaud, P. N. M. Hoang, C. Girardet, A. Meredith, A. J. Stone, Surf. Sci. **294**, 149 (1993). {9}
32. D. Schmicker, J. P. Toennies, R. Vollmer, H. Weiss, J. Chem. Phys. **95**, 9412 (1991). {9}
33. H. Weiss, Surf. Sci. **331-333**, 1453 (1995). {9}
34. J. Heidberg, H. Henseler, Surf. Sci. **427**, 439 (1999). {9}
35. J. Heidberg, M. Hustedt, J. Oppermann, P. Paszkiewicz, Surf. Sci. **352-354**, 447 (1996). {9}
36. J. Heidberg, M. Hustedt, J. Oppermann, P. Paszkiewicz, Z. Phys. Chem. **215**, 669 (2001). {9}
37. J. Heidberg, D. Meine, B. Redlich, J. Electron Spectrosc. Relat. Phenom. **64/65**, 599 (1993). {9}
38. V. Panella, J. Suzanne, P. N. M. Hoang, C. Girardet, J. Phys. I (France) **4**, 905 (1994). {9}
39. J. Heidberg, M. Kandel, D. Meine, U. Wildt, Surf. Sci. **331-333**, 1467 (1995). {9}
40. J. Heidberg, E. Kampshoff, R. Kühnemuth, O. Schönekäs, H. Stein, H. Weiss, Surf. Sci. Lett. **226**, L43 (1990). {9}
41. O. Berg, L. Quatrocci, S. K. Dunn, G. E. Ewing, J. Electron. Spect. Related Phenom. **54/55**, 981 (1990). {9}
42. H. H. Richardson, C. Baumann, G. E. Ewing, Surf. Sci. **785**, 15 (1987). {9}
43. H.-C. Chang, H. H. Richardson, G. E. Ewing, J. Chem. Phys. **89**, 7561 (1988). {9}
44. C. Noda, G. E. Ewing, Surf. Sci., **240**, 181 (1990). {9}
45. R. Disselkamp, H.-C. Chang, G. E. Ewing, Surf. Sci. **240**, 193 (1990). {9}
46. V. M. Rozenbaum, Zh. Eksp. Teor. Fiz. **99**, 1836 (1991) [Sov. Phys. JETP **72**, 1028 (1991)]. {12, 14, 18, 20, 26, 47, 59}

47. B. C. Kohin, J. Chem. Phys. **33**, 882 (1960). {12, 27, 28, 45}
48. D. A. Goodings and M. Henkelman, Can. J. Phys. **49**, 2898 (1971). {12, 27}
49. V. A. Slusarev, Yu. A. Freiman, I. N. Krupskii, and I. A. Burakhovich, Phys. Stat. Sol. B **54**, 745 (1972). {12, 27, 28}
50. J. C. Raich and N. S. Gillis, J. Chem. Phys. **66**, 846 (1977). {12}
51. F. Malder, G. van Dijk, and A. van der Avoird, Mol. Phys. **39**, 407 (1980). {12}
52. V. M. Rozenbaum and S. H. Lin, J. Chem. Phys. **112**, 9083 (2000). {12, 27, 121, 124}
53. V. M. Rozenbaum, Ukr. Fiz. Zh. **36**, 302 (1991); J. Electron. Spectr. **56**, 373 (1991). {13, 14, 18}
54. P. I. Belobrov. R. S. Gekht, and V. A. Ignatchenko, Zh. Eksp. Teor. Fiz. **84**, 1097 (1983) [Sov. Phys. JETP **57**, 636 (1983)]. {14, 66}
55. J. M. Luttinger and L. Tisza, Phys. Rev. **70**, 954 (1946); **72**, 257 (1947). {14, 40}
56. V. M. Rozenbaum and V. M. Ogenko, Fiz. Tverd. Tela (Leningrad) **26**, 1448 (1984) [Sov. Phys. Solid State **26**, 877 (1984)]. {14, 16, 25, 46, 60, 64, 66}
57. V. E. Klimenko, V. V. Kukhtin, V. M. Ogenko, and V. M. Rozenbaum, Ukr. Fiz. Zh. **35**, 1426 (1990). {16}
58. V. E. Klimenko, V. V. Kukhtin, V. M. Ogenko, and V. M. Rozenbaum, Phys. Lett. A **150**, 213 (1990). {16}
59. V. M. Rozenbaum, E. V. Artamonova, and V. M. Ogenko, Ukr. Fiz. Zh. **33**, 625 (1988). {16-18}
60. J. G. Brankov, D. M. Danchev, Physica A **144**, 128 (1987). {19}
61. V.M. Ogenko, V. M. Rozenbaum, and A. A. Chuiko, *Theory of Vibrations and Reorientations of Surface Groups of Atoms* (in Russian), Naukova dumka, Kiev, 1991. {19. 23, 25, 52, 54, 59, 87, 95, 97, 169, 173}
62. D. J. Scalapino, Y. Imry, and P. Pincus, Phys. Rev. B **11**, 2042 (1975). {21, 25, 26}
63. M. Takahashi, J. Phys. Soc. Jpn. **50**, 1854 (1981). {21, 25, 26}
64. P. I. Belobrov, V. A. Voevodin, and V. A. Ignatchenko, Zh. Eksp. Teor. Fiz. **88**, 889 (1985) [Sov. Phys. JETP **61**, 522 (1985)]. {21}
65. F. J . Dyson, Commun. Math. Phys. **12**, 91 (1969). {21}
66. Yu. M. Malozovskii and V. M. Rozenbaum, Zh. Eksp. Teor. Fiz. **98**, 265 (1990) [Sov. Phys. JETP **71**, 147 (1990)]. {21, 23, 25, 47}
67. Yu. M. Malozovsky and V. M. Rozenbaum, Physica A **175**, 127 (1991). {21, 23-26, 47}
68. A. Z. Patashinskii and V. L. Pokrovskii, Fluctuation Theory of Phase Transitions Pergamon Press, Oxford, 1979. {23}
69. N. D. Mermin and H. Wagner, Phys. Rev. Lett. **17**, 1133 (1966). {23}

70. V. M. Rozenbaum, Pis'ma Zh. Eksp. Teor. Fiz. **63**, 623 (1996) [JETP Lett. **63**, 662 (1996)]; Zh. Eksp. Teor. Fiz. **111**, 669 (1997) [JETP **84**, 368 (1997)]. {24, 40, 47, 66, 125}
71. J. Villain, R. Bidaux, J. P. Carton, and R. Conte, J. Phys. (Paris) **41**, 1263, (1980). {24}
72. E. F. Shender, Zh. Eksp. Teor. Fiz. **83**, 326 (1982) [Sov. Phys. JETP **56**, 178 (1982)]. {24}
73. D. M. Danchev, Physica A **163**, 835 (1990). {24}
74. A. F. Sadreev, Phys. Lett. A **115**, 193 (1986). {24}
75. V. M. Rozenbaum and V. M. Ogenko, Pis'ma Zh. Eksp. Teor. Fiz. **35**, 151 (1982) [JETP Letters **35**, 184 (1982)]. {25, 47, 49}
76. S. Romano, Nuovo Cimento D **9**, 409 (1987). {26, 47}
77. J. A. Goychuk, V. V. Kukhtin, and E. G. Petrov, Phys. Stat. Sol. B **149**, 55 (1988). {26}
78. I. Amdur, E. A. Mason, and J. E. Jordan, J. Chem. Phys. **27**, 527 (1957). {27, 29}
79. V. E. Klymenko and V. M. Rozenbaum, J. Chem. Phys. **110**, 5978 (1999). {27, 38}
80. I. G. Kaplan, *Theory of Molecular Interactions* (Elsevier, Amsterdam, 1986). {29}
81. V. M. Rozenbaum, Phys. Lett. A **176**, 249 (1993). {29, 31, 62, 66, 67, 71, 121}
82. P. N. M. Hoang, S. Picaud, and C. Girardet, J. Chem. Phys. **105**, 8453 (1996). {29}
83. O. G. Mouritsen and A. J. Berlinsky, Phys. Rev. Lett. **48**, 181 (1982). {33, 51}
84. A. B. Harris, O. G. Mouritsen, and A. J. Berlinsky, Can. J. Phys. **62**, 915 (1984). {33}
85. S. Tang, S. D. Mahanti, and R. K. Kalia, Phys. Rev. B **32**, 3148 (1985). {36, 46}
86. H. Y. Choi, A. B. Harris, and E. J. Mele, Phys. Rev. B **40**, 3766 (1989). {36, 46}
87. H. Shiba, Solid State Commun. **41**, 511 (1982). {36}
88. C. R. Fuselier, N. S. Gillis, and J. C. Raich, Sol. State Comm. **25**, 747 (1978). {36}
89. A. B. Harris and A. J. Berlinsky, Can. J. Phys. **57**, 1852 (1979). {37}
90. S. F. O'Shea and M. L. Klein, Chem. Phys. Lett. **66**, 381 (1979). {37, 50}
91. D. Marx, O. Opitz, P. Nielaba, and K. Binder, Phys. Rev. Lett. **70**, 2908 (1993). {37}
92. F. Y. Hansen and L. W. Bruch, Phys. Rev. B **51**, 2515 (1995). {37}
93. B. Bussery and P. E. S. Wormer, J. Chem. Phys. **99**, 1230 (1993). {39}
94. A. Wada, H. Kanamori, and S. Iwata, J. Chem. Phys. **109**, 9434 (1998). {39}

95. V. M. Rozenbaum, A. M. Mebel, and S. H. Lin, Mol. Phys. **99**, 1883 (2001). {39}
96. O. Nagai and T. Nakamura, Prog. Theor. Phys. **24**, 432 (1960). {40}
97. J. Felsteiner, Phys. Rev. Lett. **15**, 1025 (1965). {40}
98. H. Miyagi and T. Nakamura, Prog. Theor. Phys. **37**, 641 (1967). {40}
99. J. Heidberg, M. Grunwald, M. Hustedt, F. Traeger, Surf. Sci. **368**, 126 (1996). {42, 45, 62, 121, 123}
100. J. Ashkin and E. Teller, Phys. Rev. **64**, 178 (1943). {44}
101. R. J. Baxter, *Exactly Solved Models in Statistical Mechanics* (Academic Press, London, 1982). {44, 48}
102. R. V. Ditzian, J. R. Banavar, G. S. Grest, and L. P. Kadanoff, Phys. Rev. B **22**, 2542 (1980). {44}
103. P. J. Wojtowics, Phys. Rev. **116**, 32 (1959). {46}
104. T. Tsuzuki, Progr. Theor. Phys. 57, 812 (1977). {46}
105. J. V. Jose, L. P. Kadanoff, S. Kirkpatrick, and D. R. Nelson, Phys. Rev. B **16**, 1217 (1977). {47}
106. S. Elitzur, R. B. Pearson, and J. Shigemitsu, Phys. Rev. B **19**, 3698 (1979). {47}
107. V. L. Berezinskii, Zh. Eksp. Teor. Fiz. **59**, 907 (1970); **61**, 1144 (1971) [Sov. Phys. JETP **32**, 493 (1970); **34**, 610 (1971)]. {47}
108. J. M. Kosterlitz and D. J. Thouless, J. Phys. C **6**, 1181 (1973). {47}
109. V. M. Rozenbaum, Phys. Rev. B **53**, 6240 (1996). {47, 51, 52, 60, 114, 119}
110. A. N. Morozov, L. G. Grechko, V. M. Ogenko, and V. M. Rozenbaum, Fiz. Nizk. Temp. **18**, 1018 (1992) [Sov. J. Low Temp. Phys. **18**, 715 (1992); React. Kinet. Catal. Lett. **50**, 249 (1993). {47, 51}
111. S. Romano, Phys. Scripta **5**0, 326 (1994). {47}
112. T. T. Chung and J. G. Dash, Surf. Sci. **66**, 559 (1977). {51}
113. J. Eckert, W. D. Ellenson, J. B. Hastings and L. Passell, Phys. Rev. Lett. **43**, 1329 (1979). {51}
114. S. E. Roosevelt and L. W. Bruch, Phys. Rev. B **41**, 12236 (1990). {51}
115. C. Graham, J. Pierrus, and R. E. Raab, Mol. Phys. **67**, 939 (1989). {51}
116. R. D. Etters, V. Chandrasekharan, E. Uzan,and K. Kobashi, Phys. Rev. B **33**, 8615 (1986). {51}
117. C. S. Murthy, K. Singer, M. L. Klein, and I. R. McDonald, Mol. Phys. **41**, 1387 (1980). {51}
118. Z.-X. Cai, Phys. Rev. B **43**, 6163 (1991). {51}
119. P. Tarazona and E. Chason, Phys. Rev. B **39**, 7157 (1989). {51}
120. F. Y. Wu, Rev. Mod. Phys. **54**, 235 (1982). {51}
121. V. M. Rozenbaum, V. M. Ogenko, A. A. Chuiko, Usp. Fiz. Nauk **161**, No 10, 79 (1991) [Sov. Phys. Usp. **34**, 883 (1991)]. {52, 54, 59, 92, 95, 104, 114, 169}

122. V. M. Rozenbaum, Zh. Eksp. Teor. Fiz. **107**, 536 (1995) [JETP **80**, 289 (1995)]. {52, 56, 60}
123. J. Heidberg, M. Hustedt, E. Kampshoff, V. M. Rozenbaum, Surf. Sci. **427-428**, 431 (1999). {58, 71, 75-77}
124. V. M. Rozenbaum, Phys. Rev. B **51**, 1290 (1995). {62}
125. S. Prakash and C. L. Henley, Phys. Rev. B **42**, 6574 (1990). {65}
126. I. Lyuksyutov, A. Naumovets and V. Pokrovsky, *Two-Dimensional Crystals*, Academic Press, New York, 1992. {66}
127. I. G. Chernysh, I. I. Karpov, G. P. Prikhod'ko and V. M. Shai, *Physicochemical Properties of Graphite and its Compounds*, Naukova Dumka, Kiev, 1990, in Russian. {67}
128. M. L. Dekhtyar and V. M. Rozenbaum, Surf. Sci. **330**, 234 (1995). {70, 73}
129. H. Yamada and W. B. Person, J. Chem. Phys. **41**, 2478 (1964). {71, 76}
130. W. Kohn and K. M. Lau, Solid State Commun. **18**, 553 (1976). {72}
131. T. Yamaguchi, S. Yoshida, and A. Kinbara, J. Opt. Soc. Amer. **64**, 1563 (1974). {72}
132. K. Norland, A. Ames, and T. Taylor, Photogr. Sci. Eng. **14**, 295 (1970). {73, 74}
133. C. Reich, Photogr. Sci. Eng. **18**, 335 (1974). {73}
134. G. G. Dyadyusha and A. D. Kachkovskii, Ukr. Khim. Zh. **41**, 1176 (1975) [Sov. Progr. Chem. **41**, No 11, 52 (1975)]. {73}
135. G. G. Dyadyusha, V. M. Rozenbaum, and M. L. Dekhtyar, Zh. Eksp. Teor. Fiz. **100**, 1051 (1991) [Sov. Phys. JETP **73**, 581 (1991)].{73, 83, 105}
136. M. L. Dekhtyar, Dyes and Pigments **28**, 261 (1995). {74}
137. E. Kampshoff, PhD. Thesis, Hannover, 1992. {75}
138. V. M. Rozenbaum, Surf. Sci. **398**, 28 (1998). {80, 106, 175, 176, 178}
139. I. M. Lifshitz, JETP **17**, 1017, 1076 (1947) and **18**, 293 (1948); Nuovo Cimento **4**, 716 (1956). {83, 149, 152}
140. V. M. Rozenbaum and S. H. Lin, J. Chem. Phys. **110**, 5919 (1999). {84, 107, 109, 111, 116, 178}
141. S. P. Lewis, M. V. Pykhtin, E. J. Mele, A. M. Rappe, J. Chem. Phys. **108**, 1157 (1998). {86}
142. M. V. Pykhtin, S. P. Lewis, E. J. Mele, A. M. Rappe, Phys. Rev. Lett. **81**, 5940 (1998). {86}
143. B. N .J. Persson, E. Tosatti, D. Fuhrmann, G. Witte, Ch. Wöll, Phys. Rev. B **59**, 11777 (1999). {86, 124}
144. D. N. Zubarev, Usp. Fiz. Nauk **71**, 71 (1960) [Sov. Phys. Usp. **3**, 320 (I960)]. {86, 89}
145. K. Blum, *Density Matrix Theory and Application*, Plenum Press, New York (1981). {87}

146. V. M. Rozenbaum, Zh. Eksp. Teor. Fiz. **102**, 1381 (1992) [Sov. Phys. JETP **75**, 748 (1992)]. {89, 94, 101, 102, 176}
147. C. B. Harris, R. M. Shelby, and P. A. Cornelius, Phys. Rev. Lett. **38**, 1415 (1977). {89, 97, 115}
148. R. M. Shelby, C. B. Harris, and P. A. Cornelius, J. Chem. Phys. **70**, 34 (1979). {89, 90, 98, 115}
149. A. A. Makarov and V. V. Tyakht, Poverkhnost', No. 8, 18 (1988). {89, 91, 92}
150. A. I. Volokitin, Surf. Sci. **224**, 359 (1989). {89, 91, 92, 104, 176, 180}
151. B. N. J. Persson and R. Ryberg, Phys. Rev. B **32**, 3586 (1985). {90, 99, 103, 104, 115}
152. D. C. Langreth and M. Persson, Phys. Rev. B **43**, 1353 (1991). {91, 176, 180}
153. P. Nozières and C. T. De Dominicis, Phys. Rev. **178**, 1097 (1969). {91}
154. V. M. Rozenbaum, Opt. Spektrosk. **71**, 453 (1991) [Opt. Spectrosc (USSR) **71**, 263 (1991)]. {92, 104}
155. R. K. Tomas, Proc. Roy. Soc. London A **325**, 133 (1971). {92, 93}
156. A. V. Iogansen, Opt. Spektrosk. **63**, 1186 (1987) [Opt. Spectrosc (USSR) **63**, 701 (1987)]. {93, 104}
157. A. V. Ioganseen and M. Sh. Rozenberg, Zh. Strukt. Khim. **30** (1), 92 (1989) [J. Struct. Chem. (USSR) **30**, 76 (1989)]. {93, 104}
158. I. V. Kuz'menko, V. E. Klimenko, and V. M. Rozenbaum, Opt. Spektrosk. **88**, 237 (2000) [Opt. Spectrosc **88**, 201 (2000)]. {94}
159. A. V. Rakov, Proc. Phys. Inst. Acad. Sci. USSR **27**, 111 (1964). {94, 163}
160. H. A. Kramers, Physica **7**, 284 (1940). {94, 104}
161. Yu. I. Dakhnovskil, A. A. Ovchinnikov, and M. B. Semenov, Zh. Eksp. Teor.Fiz. **92**, 955 (1987) [Sov. Phys. JETP **65**, 541 (1987)]. {94}
162. W. Hayes, O. D. Jones, R. J. Elliott, and C. T. Sennett, Proc. Internal. Conf. Lattice Dynamics, Copenhagen 1963 (1965) p. 475. {95}
163. A. A. Maradudin, Solid State Phys. **18**, 273 (1966); **19**, 1 (1966). {95}
164. V. M. Rozenbaum and V. M. Ogenko, Fiz. Tverd. Tela (Leningrad) **26**, 3711 (1984) [Sov. Phys. Solid State **26**, 2236 (1984)]. {97, 98, 103, 164}
165. W. Pauli, Festschrift für A. Sommerfeld, Leipzig (1928) p. 30. {99}
166. B. Hellsing, J. Chem. Phys. **83**, 1371 (1985). {103}
167. V. M. Rozenbaum, Opt. Spektrosk. **69**, 342 (1990) [Opt. Spectrosc (USSR) **69**, 206 (1990)]. {104}
168. Z. Y. Zhang and D. C. Langreth, Phys. Rev. Lett. **59**, 2211 (1987). {105}
169. W-K. Liu, M. Hayashi, J-C. Lin, H.-C. Chang, S.-H. Lin, and J.-K. Wang, J. Chem. Phys. **106**, 5920 (1997). {105, 111, 114}
170. R. P. Chin, J. Y. Huang, Y. R. Shen, T. J. Chuang, and H. Seki, Phys. Rev. B **52**, 5985 (1995). {105}
171. J-C. Lin, K.-H. Chen, H.-C. Chang, C.-S. Tsai, C.-E. Lin, and J.-K. Wang, J. Chem. Phys. **105**, 3975 (1996). {105, 111, 113}

172. M. A. Ivanov, L. B. Kvashina, and M. A. Krivoglaz, Fiz. Tverd. Tela **7**, 2047 (1965) [Sov. Phys.-Sol. State **7**, 1652 (1966)]. {105}
173. M. A. Ivanov, M. A. Krivoglaz, D. N. Mirlin, and I. I. Reshina, Fiz. Tverd. Tela **8**, 192 (1966) [Sov. Phys.-Sol. State **8**, 150 (1966)]. {105}
174. B. N. J. Persson, F. M. Hoffman, and R. Ryberg, Phys. Rev. B **34**, 2266 (1986). {106, 116}
175. V. M. Rozenbaum, Fiz. Tverd. Tela **40**, 152 (1998) [Phys. Sol. State **40**, 136 (1998)]. {106}
176. B. N. J. Persson and R. Ryberg, Phys. Rev. B **40**, 10273 (1989). {111-114}
177. Z. Ye and P. Piercy, Phys. Rev. B **47**, 9797 (1993). {111}
178. P. Jakob and B. N. J. Persson, Phys. Rev. B **56**, 10644 (1997). {111}
179. P. Jakob, J. Chem. Phys. **108**, 5035 (1998). {111}
180. R. Honke, P. Jakob, Y. J. Chabal, A. Dvořák, S. Tausendpfund, W. Stigler, P. Pavone, A. P. Mayer, and U. Schröder, Phys. Rev. B **59**, 10996 (1999). {111}
181. H. Ueba, Progress in Surf. Sci. **55**, 115 (1997). {111}
182. P. Dumas, Y. J. Chabal, and G. S. Higashi, Phys. Rev. Lett. **65**, 1124 (1990). {113}
183. G. D. Mahan and A. A. Lucas, J. Chem. Phys. **68**, 1344 (1978). {115}
184. V. M. Rozenbaum and I. V. Kuzmenko, Surf. Rev. Lett. **5**, 965 (1998). {116, 120, 176, 177, 180, 181}
185. W. Erley and B. N. J. Persson, Surf. Sci. **218**, 494 (1989). {117, 121, 123}
186. V. M. Rozenbaum and S. H. Lin, Surf. Sci. **452**, 67 (2000). {117, 120}
187. I. V. Kuzmenko and V. M. Rozenbaum, *Proceedings of the XIV International School-Seminar on Spectroscopy of Molecules and Crystals*, Odessa, Ukraine (June 7-12) 1999, p. 200. {121, 123}
188. I. M. Lifshitz, JETP **12**, 117 (1942). {149}
189. E. W. Montroll and R. B. Potts, Phys. Rev. **100**, 525 (1955). {149}
190. E. W. Montroll and R. B. Potts, Phys. Rev. **102**, 72 (1956). {149}
191. A. A. Maradudin, P. Mazur, and E. W. Montroll, Rev. Mod. Phys. **30**, 175 (1958). {149}
192. R. L. Mössbauer, Z. Phys. **151**, 124 (1958). {149}
193. Yu. M. Kagan and Ya. A. Iosilevskii, Zh. Eksp. Teor Fiz. **42**, 259 (1962) [Sov. Phys. JETP **15**, 182 (1962)]. {149}
194. R. Brout and W. M. Visscher, Phys. Rev. Lett. **9**, 54 (1962). {149}
195. S. Takeno, Progr. Theor. Phys. (Kyoto) **29**, 191 (1963). {149}
196. A. A. Maradudin, *Theoretical and experimental aspects of the effects of point defects and disorder on the vibrations in crystals*, Solid State Phys. **18**, 273 (1966); **19**, 1 (1966). {149}
197. H. Böttger, *Principles of the Theory of Lattice Dynamics* (Academie-Verlag, Berlin, 1983). {149, 176}
198. M. Wagner, Phys. Rev. **131**, 2520 (1963). {149}

199. M. Wagner, Phys. Rev. **133**, A750 (1964). {149}
200. P. R. Ryasson and B. G. Russel, J. Phys. Chem. **79**, 1276 (1975). {159, 162, 163}
201. T. Bernstein, H. Ernst, and D. Freude, Z. phyz. Chem. **262**, 1123 (1981). {159}
202. V. M. Rozenbaum and V. M. Ogenko, Khim. Fiz. (in Russian) No 7, 972 (1983). {159, 168, 169}
203. V. M. Rozenbaum, V. M. Ogenko, and A. A. Chuiko, Teor. Eksp. Khim. **20**, 219 (1984) [Theor. Exp. Chem. (USSR) **20**, 208 (1984)]. {163}
204. D. F. Baisa, T. A. Gavrilko, V. Ya. Panchenko, Ukr. Fiz. Zh. **24**, 317 (1979). {163}
205. V. M. Rozenbaum and V. M. Ogenko, Zh. Fiz. Khim. (in Russian) **55**, 1885 (1981). {163}
206. A. G. Guzikevich, Teor. Eksp. Khim. **21**, 513 (1985) [Theor. Exp. Chem. (USSR) **21**, 489 (1985)]. {163}
207. V. M. Kontorovich, Usp. Fiz. Nauk **142**, No 2, 265 (1984). {164}
208. J. H. P. Colpa, Physica **56**, 185 (1971). {164}
209. P. W. Anderson, B. I. Halperin, and C. M. Varma, Phil. Mag. **25**, 1 (1972). {164}
210. B. P. Smolyakov and E. P. Khaimovich, Usp. Fiz. Nauk **136**, No 2, 317 (1982). {164}
211. N. D. Sokolov, Zh. Eksp. Teor Fiz. **23**, 392 (1952). {164}
212. M. V. Volkenshtein, *Molecular Biophysics* (in Russian), Nauka, Moscow, 1981. {164}
213. W. A. Phillips, Phil. Mag. **43**, 747 (1981). {164}
214. Yu. Kagan and L. A. Maksimov, Zh. Eksp. Teor Fiz. **79**, 1363 (1980) [Sov. Phys. JETP **52**, 688 (1980)]. {169, 172, 173}
215. Yu. I. Dakhnovskii, A. A. Ovchinnikov, and M. B. Semenov, Zh. Eksp. Teor Fiz. **92**, 955 (1987). {174}
216. A. I. Vainshtein, V. I. Zakharov, and V. A. Novikov, Usp. Fiz. Nauk **136**, No 4, 553 (1982). {174}
217. A. J. Leggett, S. Chakravarty, A. T. Dorsey, M. P. A. Fisher, A. Garg, and W. Zwerger, Rev. Mod. Phys. **59**, 1 (1987). {174}
218. T. Matsubara, Progr. Theor. Phys. **14**, 342 (1954). {176}
219. L. P. Kadanoff and G. Baym, *Quantum Statistical Mechanics* (Benjamin, New-York, 1962). {176}
220. L. V. Keldysh, Zh. Eksp. Teor. Fiz. **47**, 1515 (1964) [Sov. Phys. JETP **20**, 1018 (1965)]. {176}
221. D. C. Langreth, *Linear and Nonlinear Electron Transport in Solids*, eds. J. T. Devreese and V. E. van Doren (Plenum New York, 1976). {176}

Subject index*

* Brackets and braces indicate bibliography references and pages of the book where the subject items are defined, considered, or used.

Author index*

* Brackets indicate corresponding bibliography references (pages of the book on which the references are mentioned see in Bibliography).